Forms of Exclusion

Racism and Community Policing in Canada

Forms of Exclusion

Racism and Community Policing in Canada

by

David Baker

de Sitter Publications

LIBRARY AND ARCHIVES CANADA CATALOGUING IN PUBLICATION

BAKER, DAVID N. (DAVID NOEL)

FORMS OF EXCLUSION: RACISM AND COMMUNITY POLICING IN CANADA / DAVID N. BAKER.

INCLUDES INDEX.

ISBN 1-897160-21-6

1. COMMUNITY POLICING--CANADA. 2. DISCRIMINATION IN LAW ENFORCE-MENT--CANADA. 3. POLICE-COMMUNITY RELATIONS--CANADA.

HV7936.C83B35 2006   363.2'3'08900971   C2005-907684-4

Copyright © 2006 David Baker

All rights are reserved. No part of this publication may be reproduced, translated, stored in a retrieval system, or transmitted in any form or by any means, electronic, mechanical, photocopying, recording or otherwise, without prior written permission from the publisher.

Authorization to photocopy items for internal or personal use is granted by the publisher provided that the appropriate fees are paid directly to de Sitter Publications.

Fees are subject to change.

Cover design: de Sitter Publications

de Sitter Publications
http://www.desitterpublications.com
sales@desitterpublications.com

Table of Contents

Acknowledgements vii

Preface ix

Chapter 1
    Principals and Paradigms 1

Chapter 2
    Personal and Up Close: Young Black Men Speak on Crime, Race and Community Policing 31

Chapter 3
    Criminalization: Racializing Crime 41

Chapter 4
    Race, Citizenship, and Exclusion 67

Chapter 5
    The Police in Community Policing: A Black Viewpoint 85

Chapter 6
    Afterwards 101

References 113

Index 121

Acknowledgments

I wish to thank a number of colleagues and friends who influenced this book on racism and forms of exclusion. At the level of support on a day-to-day basis, Georgia Graham was a constant source of inspiration. She encouraged me in her own unassuming way. I especially thank the students at York University, Humber College, and the Jane/Finch community members who participated in the focus groups. I thank all of these people for putting up with my demands. I treasure them as friends and colleagues. To all community members who assisted me in countless ways: thanks for your time and information. You have helped to guide this work over the years.

Finally, I would like to thank de Sitter Publications and its staff for acquiring this manuscript, for allowing me the necessary creative freedom and flexibility to pull this book off, and for being so easy to work with despite hectic schedules and busy agendas.

Preface

This book is for people concerned with the problems of racial injustice in society. In particular, it deals with state systems of control and regulation. Readers, that is, students and teachers of criminology, sociology, law, policy makers, and criminal justice professionals, are persuaded that minority ethnic groups suffer substantial inequalities. Over the past two decades there has been a proliferation of research on policing and race relations, a significant proportion of which has been concerned with community policing. The preponderance of written materials on this issue has been impressive. Indeed, it would not be an exaggeration to say that the concept of community policing is now ubiquitous in the humanities and social sciences, being invoked not just by the political scientist and the sociologist, but also by the philosopher and the literary critic. No book can hope to do justice to the breadth of material and certainly I would not claim any such pretension myself. My aim in this book is considerably more limited. I attempt to return to some of these original sources, to inquire analytically into the fundamental premise and include my own reflections, realizing that no single theory or intellectual positioning is capable of comprehending the phenomenal complexity of even a moment of experience, assuming the possibility of locating and isolating that elusive moment. Admittedly, more research needs to be specifically directed at race issues. I argue that all criminological research should encompass dimensions of racial differentiation and gender differentiation. Thus, engaging or actually doing both is difficult and painful. This book acknowledges the pain and the problems, and aims to assist research on race issues by opening up methodological debates and by presenting a range of theoretical options.

In the process of my research I have encountered many difficulties. Difficulties have, in part, centered on the fragmented and contentious nature of available literature. My library search led me to pursue different disciplines: sociology, cultural studies, criminology, history, politics, social policy, and geography to name a few. In many of these interrelated disciplines, the presence of racism and the concomitant problems, which warrant explanation are well-known, so that theoretical disputes can be fruitful and progressive. Interestingly, when it comes to racism and criminology, I found more attention was given to defining the problem rather than to productive theorizing. It is no secret that in criminology the dominant and "privileged" theoretical perspective is the administrative perspective.

After analyzing the contributions of theorists who write about whether or not there is a problem of black crime or black criminalization, or policy researchers who argue about the existence of discrimination in the criminal justice system, there is a real sense in which both are locked into the Black/White dichotomy. Further, theoretical analyses are criticized for being "speculative," unworried by empirical data, whereas policy research is questioned on the grounds that it endlessly seeks data ultimately fruitless in the absence of theoretical postulates. A criminological understanding of racism seems not to have progressed beyond the polarized debate of the 1980s and the Black/White dichotomy. However, more recently there have been significant advances in the official acknowledgement of racism in the criminal justice system.

My guiding perspective has, of course, been a sociological one, and I have approached the subject matter acknowledging such traditional concerns as the meaning of the term "community," the epistemic questions of its relation to science, and the problems raised by the notion of "community policing." Nevertheless, in what I hope will be regarded as fertile eclecticism, I have not hesitated to draw on material from other disciplines where necessary. My belief is that this has helped to illustrate some of the more abstract points and has thereby periodically given the discussion an applied dimension.

It is particularly timely to do a racial analysis of community policing today. A number of local, national, and international events have supported "law and order" trends and their connections to racism, sexism, and classism within the "right-wing backlash." A cursory reading of works on racism during the post-Civil Rights Movement era reveals a general consensus that race is not salient. But when race does matter, it is situated within a discourse of criminality. The media, one of the most powerful sites through which societal images are shaped, play a pivotal role in the perpetuation of racism. Mainstream media consistently depict liberalized images of a society that try to move beyond race-related problems toward multiculturalism and the "new" talk of diversity. In light of these developments, race-based identities and policies discourses seem regressive, at a time when race permeate almost every facet of Canadian life.

In late capitalism, an era marked by capitalist expansion and globalization, Marxist perspectives have been discredited internationally in light of the fall of the transitional socialist governments in Eastern Europe. Ethnicity/race only exists as long as it does not interfere or determine consumption habits. In the post-NAFTA era, capital has become more and more flexible in crossing national boundaries while, at the same time, nationalism in the form of "ethnic pride" and "ethnic cleansing" seems to be sweeping many societies. In such discourse, race is restricted to a cosmetic function at a point in history when many cities in Canada, such as Toronto are more diverse racially than ever. It may seem, at times, that the motive of Black people's emancipatory struggles has been to make race and ethnicity unimportant. Arguably, this has arrived in the late 1990s, as one takes a cursory look at images presented by Hollywood and the mass media. Yet, some argue that Blacks are dissatisfied with such representations and suspicious of the underlying motives. A critical approach to seeing and reading these images may reveal that not only do policies and laws serve a historically oppressive interest, but these "new" racial views and attitudes thrive on particular misconceptions about empowerment. Somehow, the fate of Black people's social struggles, including social justice in Western society, became inextricably tied to the end of racism.

Anti-racism as a strategy to resist racism is counter-productive because the discourse keeps Whites dominant and at the center of the debate, in essence reinforcing White supremacy. In most of these debates, whether it is within a particular site, for example the criminal justice system or policy issues, exclusion automatically means oppression, while inclusion means empowerment. Affirmative Action policies or programs easily attest to this exclusionary project. Modern day race policy rushes ahead by applying a simplistic logic that exclusion is bad and inclusion is good. The equality achieved with this logic is at best asocial, because racial economic equality certainly does not exist. With economic disparities in place, integration is simply a prelude to assimilation.

Some Black thinkers continue to hold the Civil Rights Movement as an ideal, which, although it may have outlived its utility, has not been discarded by Whites. It is a point of reference for conservative thinkers to demonstrate how far race relations have progressed since the 1960s. As Omi and Winant (1987) assert, the old recipes for racial equality, which involved the creation of a "color-blind" society, have been transformed into formulas for the maintenance of racial inequality. This is all too clear when one looks at the "law and order" debates of the past decade. This was a time that brought forth the Rodney King trial and its aftermath in Los Angeles. Also at this time, Canada experienced the shooting of unarmed Black men by the police, the death of Georgina Leimonis in the Just Desserts Cadi robbery, and the shooting of Police Officer Todd Baylis in Toronto. All of this occurred against the backdrop of the call for "law and order." In Canada, the Federal Government strongly pushed a "law and order" agenda. On June 15, 1995, the "Danger to the Public" provision of the *Immigration Act* received Royal Assent. This Act removes the right of refugee claimants to seek protection in Canada and the rights of permanent residents of Canada to appeal their deportation order to the Immigration Appeal Division of the Immigration and Refugee Board. Further, Bill C-17 *Criminal Law Improvement Act, 1996*, gives the police expanded powers of arrest, the power to set terms of judicial interim release, and further crime exemption for police and police agents.

This discourse now serves to legitimize the popular and prevailing racial attitudes of the day. Essentially, neo-liberal thinkers have allowed an outdated conservative agenda to succeed. That is, the issue of race has been diffused and the public is weary of racism. However, the public keeps a keen eye out for "reverse racism," or any other exclusionary practices exercised by ethnic groups. Thus, the construction and ensuing dissolution of race serves particular interests. Today's integrative race policies or color-blind theory and attitudes (clearly an element in Canadian multiculturalism and community policing) are in actuality assimilative policies designed to eliminate or diffuse difference. Such attitudes are structured to serve the interests of late capitalism by allowing capitalism the stability it needs during a transitional or contradictory phase.

Social perceptions about the law and the economy influence how resources are distributed in certain sectors of society. In this sense, as societies move through the twenty-first century and embrace the global economy, race is once again been pushed aside and the economic, intellectual, and social investments of the people in these racial groups are diversified.

The logic of late capitalism produces a racial discourse, which allows ethnicity to flourish as long as it is dispersed and not organized. Here, Black people are treated as packages and sold self-determination through the equal opportunity to consume. Concurrently, Black communities are encouraged to accept social peace at the cost of social justice. It is within this logic that this book argues that community policing serves to maintain the status quo, rather than empowering Black communities. Further, this book serves as a documentation of the forms of exclusion through a discourse of community policing.

Chapter one presents an overview of some of the principal approaches that have been taken to the study of racism and criminology. This chapter takes the view that although traditional approaches highlight the problems of racism in society, it is not sufficient in conceptualizing racism today. This chapter advances critical race theory and critical criminology, insofar as it does not limit the scope of inquiry to state definitions of crime and prefers to include issues of social harm and social justice. The chapter also takes a critical look at the *Report of the Commission on Systemic Racism in the Ontario Criminal Justice System*. In so doing, it seeks to expose and oppose domination rather than be complicit in its recreation.

Chapter two elaborates upon the methodological issues raised in chapter one. Specifically, this chapter looks at young, law-abiding Black University and College male students. Utilizing focus groups, this chapter captures a specific point of view on crime, race, and community policing that is so often overlooked.

Chapter three examines the media images of Blackness. This favorable image serves to conceal rather than reveal the forms that race/racism takes in the discourse on community policing at particular moments of the law and order "crises." This chapter then presents the works of Antonio Gramsci, Livy Visano, and others that caution us about hegemony, law, consent, and ideologies that produce a particular frame, that so-called reasoning becomes the ideological reason in the service of the state.

Chapter four explains the utility of the concept of citizenship in understanding both the ideological and material conditions under which

Black people live their lives in Canada. It is argued that exclusion manifests itself through the denial of legal citizenship through immigration policies.

Chapter five looks at the role of the police in community policing from a Black viewpoint. I argue that community policing is another form of societal discipline.

Lastly, chapter six highlights and problematizes some of the primary concerns and forms in the discourse on racism and community policing.

Chapter 1

Principals and Paradigms

Introduction

The aim of this chapter is to present an overview of the principal approaches used in the criminological study of race and highlight some of the most important concepts that have been generated and some of the most valuable insights that have been produced. Some of the major controversies and unresolved contradictions that have arisen within, and between, different criminological perspectives will be discussed. The intention is not to try to provide an exhaustive summary of the research on race and racism in crime and criminal justice. The purpose, rather, is to focus on the theoretical allegiances and assumptions that underpin various criminological engagements with race issues addressed through research, which can generally be categorized under three main headings:

1. Race and criminality
2. Race, racism, and criminal justice
3. Racism and criminalization

These issues have been dealt with by the major criminology orientations, following the descriptions of criminological paradigms used in reviews of the development of British criminology (Rock 1988). Although this chapter is informed by British criminological practice, it has relevance to Canadian and American criminology as well.

Theoretical Overview

Much of the research on race, crime, and criminal justice has started from supposed facts and figures concerning the Black and minority presence in penal populations. The racial composition of penal and arrest populations is derived from prison statistics, where ethnic information is recorded in the annual census and the police record the ethnic origin of all those arrested, cautioned, or referred for protection. Mainstream criminologists have used official statistics as a starting point for their investigation of high crime rates among Blacks. Mainstream criminologists find explanations for high crime rates among the Black population in elements of the culture of the so-called "underclass," "ghetto," or poor/low-income housing areas. In these accounts, criminal behavior is a conscious choice by members of social groups who are hostile to or disrespectful of authority in general and of the police in particular. They lack a healthy work ethic, they lack law-abiding parental role models, and they are involved with drugs.

While mainstream criminologists stress the element of choice in the criminality of Black ghettos/low-income areas, broader sociological accounts describe the culture of dependence resulting from the reliance on welfare benefits, and the effects of affirmative action and employment equity policies in enabling the "brightest and best" of Black community members to leave the ghetto or low-income areas (Auletta 1982). These versions of underclass theory are reminiscent of the early Chicago School criminologists, who described inner-city areas as zones of transition illustrated by family disorganization and weak informal social controls. Similar explanations have been put forward regarding crime among the migrant workers and colonial immigrants of Europe (Junger 1989). Essentially, traditional approaches argue that criminogenic factors of Black and migrant cultures are taken as a given, and it is their role in Black criminality that is being uncovered, rather than the causes of these (presumed) cultural aspects themselves.

The emergence of a Black middle class has been taken as evidence that racism is not a significant factor in the production of crime-prone Black and minority subcultures. Three specific types of criminology consistently engage the issue of race: (a) administrative criminology, which is orientated primarily to issues of criminal justice policy; (b) radical criminology, with its aim of illuminating the fundamental nature of the crime problem; and (c) critical criminology, with its aim of analyzing the nature of the state and, in particular, the exercise of state power (Hudson 1994). Interestingly, wherever there are findings of a hostility toward police, and/or a lack of enthusiasm for finding work, neo-conservatives blame those social policies (affirmative action/employment equity, generous welfare benefits) that were designed to help improve the chances of ghetto/low-income residents of finding work or avoiding extreme poverty, rather than the structural unemployment and racism that necessitated the adoption of such policies in the first place (Murry 1984). The notion of "mainstream" criminology refers to criminology that was not, or is not, influenced by the paradigm shift away from positivist criminology developed by the labeling theorists and interactions of the 1960s and 1970s (Young 1988). As Rock (1988), Reiner (1988), and Young (1988) remind us, mainstream criminology is not considered mainstream in that it is necessarily most numerous or most intellectually dominant; rather, it has not departed from the mainstream positivist-empiricist orientation of applied social science. Although mainstream criminology may be on the theoretical margins, it continues to have influence.

Mainstream criminology is not concerned with the role of the state in producing crime or the social reactions that help produce criminal identities. And, it fails to acknowledge crime and criminal justice as contingent outcomes of social-political configurations (Hudson 1994). Because of its lack of concern with structural factors, one would not expect mainstream criminology to have contributed much to the understanding of racism. Interventions on race issues have been addressed under the rubric of race/criminality questions: What, if any, are the differences between Black and White criminals? What are the predictors of criminality among Black and other minority ethnic groups? As Reiner (1989) points out, "mainstream criminology poses its questions and therefore produces its explanations at individual and cultural levels of analysis but not at structural levels" (p.5) Further, Hudson (1994) reminds us that "mainstream criminology uses established definitions of crimes, and proceeds by established methods and theories" (p.3).

The uncritical acceptance of official criminal justice and law enforcement statistics as accurate indicators of participation in crime is the hallmark of mainstream criminology. One of the significant roadblocks to an informal discussion about crime and race is the perpetuation of detractors. Detractors are statements or propositions about crime that are discussed in a vacuum, divorced from their contexts. James Q. Wilson's discussion of the relationship between White racism and Black crime provides an interesting example of this phenomenon. Others, for example

Hacker (1992) and Cole (1999) have defended the argument that White racism is at the root of the problems faced by Black Americans as well. I would like to draw attention to the work of Russell (1998). What follows are summaries of her readings of Wilson's work on race and crime.

Wilson's basic thesis, according to Russell (1998), is that if Blacks would stop committing so much crime there would not be so much White racism. His thesis, developed in subsequent writings, suggests that "White racism and White fear of Black and Latino men are justified because Black and Latino men have high rates of crime" (p.125). Wilson contends it is fear, not racism, which accounts for the negative perceptions that White people have of Black and Latino men. Wilson argues that "fear can produce behavior that is indistinguishable from racism" (p.125). His tacit conclusion is that the current level of White racism is acceptable, so long as it coexists with the current level of Black and Latino crime. At first glance, Wilson's argument sounds vaguely tenable or at least difficult to dismiss. However, a careful consideration of his underlying premise indicates that his thesis raises more questions than answers.

Fundamentally, Wilson suggests the following one-dimensional relationship:

Black crime rates ⟶ **White racism**

Two major assumptions form the foundation of Wilson's hypothesis. First, that Black crime rate is the primary source of White racism. Second, solving the Black crime problem rests primarily with the Black communities. According to Wilson, "the best way to reduce racism, real or imagined, is to reduce the Black crime rate to equal the White crime rate" (as quoted in Russell 1998:128). He asserts that Black men offend at a rate of six to eight times greater than Whites. Accordingly, it is reasonable to expect that White racism will persist until Blacks and Whites offend at an equal rate. Awaiting such a drop in the Black crime rate is neither the best nor the quickest way to reduce White racism. Ignoring the interconnection between crime, poverty, and education, Wilson commands Blacks to rise above their circumstances before they can ask for a reduction in White racism. Indeed, this is a tall order.

Not only does Wilson imply that the Black crime rate is the sole source of White racism, he also places the onus of eradicating White racism upon Blacks. Even if the Black crime rate was reduced to equal the White crime rate, how would this affect the amount of White racism? Is Wilson suggesting that if Black and White crime rates were equal, White racism would wither away or decline substantially? Wilson provides neither theoretical nor empirical support for this sweeping assertion. However, this book draws upon an interesting configuration: (a) White racism surfaced only in response to high Black crime rates; (b) Whites have a passive role in the Black crime/White racism dynamic. Simply put, he blames Blacks for White racism and Blacks unfairly use racism as an excuse for criminal activity.

As Wilson speculates that the Racism that is "imagined" by Blacks will disappear if the Black crime rate declines, he fails to consider the definition of "imagined" racism. One is left to guess that this racism only exists in the minds of Blacks. How is it that a reduction in the Black crime rate would result in a reduction in imagined racism? Here, too, he fails to adhere to definition of "imagined" racism. Rather than holding Whites accountable for their racism, Wilson allows them to claim victim status. They are victimized by Black crime. As Russell (1998) points out, "Wilson allows the blame for White racism to be placed entirely on Black shoulders. Yet, he charges that Blacks unfairly place all the blame for the Black crime rate on Whites" (p.129). This line of reasoning encourages us to think along racially segregated tracks about crime and other societal problems. His arguments suggest that Blacks are responsible for Black crime and that White racism is part of the larger societal racial finger pointing.[1]

The overemphasis of research on Black crime makes it difficult to see that race and crime is not synonymous with Blacks and crime. More must be done to present the public with an accurate racial picture of crime, including the effects of, and the data related to, White crime. Rather than investigating the social processes that contribute to the production of statistics; that is, the decision-making activities of legislators in criminalizing behaviors, mainstream criminologist like Wilson take the outcome of these processes as their starting point for the construction of ideological causes of crime rates among sections of the population. Perhaps we should examine, specifically, the consequences of this theoretical framework.

Administrative Criminology

During the 1980s, administrative criminology, a term which came into common criminological currency, describes the work of criminologists engaged in applied research, that was aimed primarily at assisting criminal justice and penal system professionals in policy development and decision-making (Hudson 1994). The objectives of this paradigm are effectiveness, efficiency, and matching practices to policies. Conceptually grand theorizing is decreased. Much of administrative criminology is commissioned directly or indirectly by government or well-established non-governmental agencies.[2] It therefore arises from practice and policy concerns rather than the concerns of scholarly curiosity, theoretical debate, and intellectual growth

Although administrative criminology encompasses both crime and criminal justice, it has, as Cohen (1981) predicted, largely concentrated on criminal justice, and an exclusive concern with the operation of the system, rather than with the causes of people coming into the system or their involvement with the system as victims, suspects, or defendants. When dealing with issues of race, much of administrative criminology has been focused on whether or not criminal justice and penal system agencies and processes discriminate against people on account of their skin color or ethnic affiliation.

Unlike mainstream criminology, administrative criminology understands that criminal justice records and decisions are the outcomes of social processes. It does not concern itself so much with the morality of the outcomes, but with whether the outcomes can be justified by proper adherence to processes and procedures (Hudson 1994). The response to this predicament has been a recurring ambivalence that helps explain the volatile and contradictory character of recent crime-control policies. The efforts of administrative criminology have been unwavering in the production of predictive instruments to aid decision-makers, and monitoring instruments to curb discretion.

Administrative criminology has concentrated on discretion and on individual processes or agencies, rather than on seeing criminal justice and the penal system as a whole. It has produced statistical enquiries, which, even as they become more methodologically sophisticated, continue to produce findings that contradict each other and contradict lived experience, particularly Black experience. It has contributed to more effective management of various agencies and processes in the administration of justice but has contributed little to racial justice or to criminological understanding.

Radical Criminology

Administrative criminology has been influential with policy makers and practitioners. Radical criminology has been most influential with academic criminologists. It has brought about an understanding of what is often called "Black crime." The central claim of Radical criminology is that crime, and the state's reaction to crime, can only be understood in the context of a full sociological framework. As such, Hudson (1994) argues "that the nature, extent and location of crime and the nature, extent and location of control can only be explained with reference to the material and ideological relationships that exist within a social formation" (p.12). To put it more precisely, Hudson's (1994) claim is that crime in a capitalist society can only be understood in the context of the class relationships. Such a claim fits well within Young's (1988) definition of Radical criminology, which states

> that part of the discipline, which sees the causes of crime as being at core the class, and patriarchal relations endemic to our social order and which sees fundamental changes as necessary to reduce criminality. This is politically at base socialist, libertarian/anarchist, socialist or radical feminist. It quarrels amongst itself- as such a radical mix has throughout history- but it is quite distinct from those parts of the discipline which see crime as a marginal phenomenon solvable with technical adjustments by control agencies which seems all right and in need of no fundamental change (P.160)

It is clear, then, that Radical criminology sees a high incidence of crime as inevitable in a society characterized by gross inequalities in wealth and opportunity. It merges together the basics of Merton's (1964) version of anomie theory and Cloward and Ohlin's (1960) opportunity theory.

Given their understanding of the political economy of crime, radical criminologists maintain that there is bias in the processes of law enforcement and criminal justice. They view statistics of Black participation in crime, law enforcement, and criminal justice with skepticism (Hudson 1994), but also expect statistics to reflect some real and significant Black criminality. Radical criminology, then, does not ask whether race/crime statistics are derived from "Black crime" or from racist law enforcement and criminal justice, but expect them to reflect both (Hudson 1994). Similarly, research that goes beyond a simple search for racist attitudes among police can provide some explanation of how such attitudes arise and are sustained.

The research conducted by Brown and Willis (1985), Lewis (1989), and the Commission of Inquiry into Systemic Racism (1995), over the past three decades, has consistently found there are police officers who have racist attitudes. This scholarship has convincingly argued that a "canteen culture" (Fielding 1988) carries over from joking with fellow officers to encounters with the public (McNulty 1994). The belief is that a racist police culture exists because recruits share the racist culture of the group from which the majority is drawn—the macho, authoritarian, prejudiced working class—and such attitudes are reinforced not only by socialization processes upon entering the force, but also by the confrontational situations that occur between police and Black people on the job (Jefferson 1988; Reiner 1985; Vincent 1990).

Critical Criminology

The goals of critical criminology/critical sociology are to demonstrate the precise ideological constructs deployed at particular historical moments, and the investigation of ideological shifts with a focus on the drift to law and order (Hall 1980). Gilroy and Sim (1985) reminds us that this involves

> moving from securing compliance by rewards to imposing compliance by repression and involves the criminalization of various sub-groups of the disadvantaged, and their marginalization by legal regulation as well as by economic privation. We can see this in the division of the unemployed into scroungers and genuine claimants,

the restriction of rights to housing benefits etc. Black people are, therefore, not the only group to be stigmatized as "the enemy within" in the ideological drive to blame sub-sections of the powerless for their own predicament. (P.20)

There is nothing new about casting sub-sections of the poor as dangerous and/or undeserving and the moral panic about law and order. The law and order "crisis,"[3] argues that "race" is a common element, a concept that joins together the various elements of the presentation of "the crisis," as identified by Solomos, Findlay, Jones, and Gilroy (1982). Thus, the desire emphasis by critical sociology on Blacks as a particular focus in terms of the criminalization of (potentially) disaffected sub-groups of the powerless, becomes a study of the character and dynamics of what has been termed "New Racism" (Barker 1981).

The fundamental focus for critical criminologists is policing as a method of maintaining order, whereas administrative and radical criminologists have been more concerned with responses to crime. Thus, critical criminologists advance a position that coercive or paramilitary policing marks the abandonment of concern with crime in favor of the containment of disorder. It is a position that shows how perceptions of certain groups of people, who are marginalized and excluded, are used as evidence that containment tactics are necessary (Cashmore and McCauglin 1991; Jefferson 1990). Some researchers (Lea and Young 1984) would be tempted to lump the crisis of crime and disorder together, and argue for more realistic policy responses. Within this Marxist analysis much more is lost than gained in understanding the issues of race. Other critical criminologists (Hudson 1994) demonstrate how a widespread belief in, and fear of, crisis has been used to justify increased police resources and to demonstrate how the "collapse of basis for consensus policing" reverts to an increased "drift towards 'military' policing methods" (Clarke 1987).

Like radical criminology, critical criminology offers insights into some of the contradictory evidence produced by administrative criminology and explains some of the apparent contradictions. Critical criminology does not limit the scope of inquiry to definitions of crime and prefers to include issues of social harm and social justice. This approach criticizes the existing system of criminal justice as reflecting and perpetuating forms of exclusion and domination that include class, gender, race/ethnicity, and heterosexism. Critical criminology seeks to expose and oppose domination rather than be complicit in its perpetuation. It is an approach where creative and cooperative solutions are sought instead of more repressive, tougher measures of crime policies to secure an unjust social order. Critical theorists have also been wary of the faith administrative criminologists have in due process and the concept of "just deserts." Their appreciation of criminal justice as a homogenizing filter has led to incorporation of the challenge of critical legal theorists to law as a system that treats all fairly. Fitzpatrick (1987) sees the law as resting on assumptions such as procedural equality, free will, and general protection of rights that express White, male, and middle-class standpoints.

Thus far, the concept of community policing has not been adequately defined. There is much disagreement among criminologists as to what, exactly, constitutes community policing and the implications therein for communities. As noted previously, the focus of this book is not to define community policing, but rather the elusiveness of the concept. Nonetheless, according to Murphy (1988),

> [t]he wholesale endorsement of community as both the means and the end of community policing projects and programs presupposes some agreement about the conceptual and empirical validity of community as an identifiable and viable concept. While ideologically appealing, the image of community used in much of the literature is often nostalgic, consensual, geographically limited, and value laden. Police experience to date suggests a more realistic and perhaps useful conception of community as that of a community of interests, requiring some degree of mutual collaboration and agreement. This more limited, yet more empirically precise, conception of community, while less ideologically appealing, may more accurately reflect the reality of the urban policing environment and encourage community policing programs to espouse more modest and achievable objectives. (Pp.392-393)

However, Visano (1994) cautions us that "the concept of a community has been appropriated ideologically by the state to legitimate decisions, preserve privilege and maintain authority relations" (p.195). Such an appropriation facilitates the standardized, homogenized, and universalized discourse on community policing. It is a discourse that violates and silences the plurality of voices and histories, by denying and flattening contexts and diversities. This phenomenon can be traced to the emergence of new strategies and techniques, what Garland (1996) calls "responsibilization strategy."[4]

Community policing is of particular importance when dealing with issues of race or racism. This focus allows victims of racism to engage in a discourse of criminalization that further privileges the dominant ethos; that is, that race and crime are connected in some form in racially stratified societies. Given the interconnectedness of crime and race, the state alone cannot effectively be responsible for preventing and controlling crime and racism. All citizens must be made to recognize that they too have a responsibility in this regard, and must be persuaded to change their practices in order to reduce criminal opportunities and increase informal controls. The questions to be asked, then, are whether these approaches are sufficient, and, if so, do they lead to racial and social justice? What follows is a critical examination of state discourse on race and crime.

Official Publication

There is a particular British legacy that still exists in Canada within the context of the administration of law and order. That legacy is in the use of Royal Commissions by the British as an administrative strategy for bypassing legal processes to establish a form of unlawful jurisdiction. In writing about his studies on Royal Commissions and Departmental Committees, Cartwright (1975) reminds us that "this development was contingent on the separation of parliamentary political offices for ministers and administrative permanent offices for civil servants" (p.35). Such separation becomes a celebrated mode of inquiry. Indeed, as Burton and Carlen (1979) highlight, "they touched with one hand the ancient machinery of forensic inquiry, with the other hand the new method of an inductive, experimental science" (p.4). At such moments in British history, what took place were efficient state apparatuses founded upon empirical knowledge.

The questions of pauperism and poor-law administration, of crime and penal administration of pestilence and sanitary legislation, and of the evils attendant on excessive manufacturing labor, are conspicuous instances of the effects of commissions of inquiry in reversing every main principle, on almost every assumed chief elementary fact, on which the general public, parliamentary committees, and leading statesmen were prepared to legislate (Chadwick 1937:54, as cited by Burton and Carlen 1979). This knowledge and its attendant institutionalization into state practices were requirements of the ascendant capitalist class to control the social contradictions produced by an unstable and potentially revolutionary situation. The principle of order dominates legal and administrative forms as advocated by the radical bourgeoisie. As Burton and Carlen (1979) note, the "administrative goal was one of calculated intervention to keep structural contradictions under control so that economic interests could be systematically pursued" (p.5). At this moment in British history, it is useful to highlight the use of utilitarianism. Parekh (1974) explains: "Benthamite utilitarian benevolence did not extend to the working and impoverished classes who had neither time nor ability to develop social motives but were doomed to self interest" (p.13).

Within the British tradition, inquiries or commissions functioned as tools that provided and publicly propagated knowledge of social conditions that would shape the technology of social engineering. As such, the legacy still exists in Canada today insofar as inquiries or commissions have a clearly dual function of not only creating information, but also manipulating its popular reception. It is the result of a British historical resolution of the tensions between forms of government in a liberal order.

Canadians concentrate on official publications. For example, The Commission on Systemic Racism in the Ontario Criminal Justice System (hereafter called the *Cole-Gittens* report) is a representative sampling of voluminous reports produced by vast government publishing machines. Calls for public inquiries are usually prompted by tragic events. Victims want explanations and redress, while also satisfying the broader interest in uncovering systemic failures and preventing future tragedies. The subject matter covered by these investigative committees varies widely. They have examined issues ranging from contaminated blood to mining accidents.

However, a substantial portion concerns the administration of law.

Inquiries are ad hoc investigative committees that are set up accountable to the prerogative or conventional powers of the government ministers, under the aegis of the *Public Inquiries Act* (1971). Royal Commissions are established nominally by the crown to investigate and report upon specific matters defined in their term of reference. As Burton and Carlen (1979) state, "within these echelons of knowing subjects the judiciary is particularly well represented. Law lords, judges and lawyers are more than twice as likely to chair investigative committees than any other groups, usually academics and businessmen" (p.1). Foucault's (1991) notion of governmentality has significance here because it suggests alternative ways of thinking about the activity of politics. As Rose (1993) points out,

> [t]he forms of power that subject us, the system of rules that administer us, the types of authority that master us—do not find their principle of coherence in a State nor do they answer to a logic of oppression or domination or the other constitutive oppositions of liberal political philosophy—least of all, its ways of dividing the political from the non-political. The force field with which we are confronted in our present is made up of a multiplicity of interlocking apparatuses for the programming of this or that dimension of life, apparatuses that cannot be understood according to a polarization of public and private or state and civil society. (P.286)

The *Cole-Gittens* Report[5] demonstrates this all too well. Members of this Commission were appointed by the Attorney General with the help of the respective civil service departments concerned with specific topics of inquiry. They are considered lay experts in the fields of knowledge relevant to the problems in the administration of justice. This legal presence is considerably higher than departmental committees and provides the reports with their quasi-judicial character. Although formally non-judicial, some committees have the authority to mandate witnesses to give evidence (as in the Somalia Inquiry) and have at their (the Commission's) discretion the right to sit in public or private, and to publish or not publish any minutes of evidence or information in their findings.

The defining characteristics of investigative committees, including their expertise and their public, advisory, and ad hoc nature, are the foundation for their claims to impartiality and disinterestedness and are the characteristics upon which their existence is founded and protected. Formally, they are neither judicial nor administrative but occupy a consultative space that is technically external to both. This is particularly appropriate, as Burton and Carlen (1979) point out, "when the committee is not determining legislative policy but is investigating the activities of state functionaries" (p.2). However, with the help of the civil service, these honorary state functionaries produce reports that receive the state's unrehearsed acceptance and are generally ordained as officially recognized discourses.

A set of popular criticism and arguments about investigative commissions and committees can rightly be made. For example, these criticisms range from a tactical device to defray government activity to postponing legislative or other actions while simultaneously demonstrating that particular problems are under administrative review and control. Time is important but not sufficient in explaining the political relevance of "official publications." Burton and Carlen (1979) espouse this clearly in their work on "official discourse":

> Moreover the recommendations of a report (when not "whitewashed"), being advisory, can be and frequently are ignored. The potentially inconsequential outcome of a report is not conducive to restoring public confidence. Again, as the research role of these committees should be unnecessary given the development of state professionalism, the reports perform 'merely' rhetorical function. Such arguments are usually countered in the literature by pointing out those texts that have been influential, have been implemented and were dependant on lay expertise. (P.6)

The particular type of investigative committees/commissions, which are of concern here, grapple with problems in the administration of law and public order. These investigative committees will be theorized in a specific manner. Clearly, their interest is not to evaluate the direct legislative and administrative consequences of official texts, as Burton and Carlen

(1979) reminds us these documents represent a system, of which "the intellectual collusion whereby selected, frequently judicial intelligentsia, transmit forms of knowledge into political practices" (p.7). This process is one that facilitates official arguments and replenishes both established and novel modes of knowing and forms of reasoning. Further, by linking state functionaries with lay intelligentsia, official discourses on law and order become a part of the constant renewal of "hegemonic domination."[6] This practice is one amongst many in the process of reproducing specific ideological social relations. This form of intellectual collusion is a technique of discursive incorporation through which legitimacy crisis are repaired and the reforms they engender are publicly presented.

The interventionist phase of the capitalist state by overtly re-politicizing economic relations has concomitantly increased the degree of ideological control required for the reproduction of the total social formation as suggested by Habermas (1976) and Hall, Critcher, Clarke, Jefferson, and Roberts (1978). Hegemony and legitimacy crises in the interventionist era represent a phenomenal reaction to the state's inability to control the effects of the economic class struggle during a period of restructuring of capitalist relations. Correctly placing the ideological conflict of law and order debates surrounding matters such as racism, as elements within the hegemonic shifts accompanying changes in British capitalism, Hall *et al.* (1978) argue that

> [a]ny profound restructuring of the inner organization and composition of capitalist relations—such as characterized the long transition from laissez-faire to monopoly, or the more intense section of this where British capitalism found itself—requires and precipitates a consequent "recomposition" of the whole social and ideological integument of the social formation. (P.225)

The shifts in hegemonic practices considered here are reactions to forms of crises, which result from the restructuring of capitalism into corporatist structures. Such a re-composition of capitalism has resulted in an increased centralization, concentration, and internationalization of capital, and a consequent revolutionizing of the labor process. In general terms, the state reaction toward this conflict—everything from racism to transformations in sexual morality—has been to steadily increase the coercive elements of hegemonic control. Hall (1978) asserts the following:

> The mobilization of legal instruments against labor, political dissent and alternative life style, all seemed to be aimed at the same general purpose: to bring about by fiat what could no longer be won by consent—the disciplined society...The growth of political dissent from the mid-1960s onwards, then the resumption of a more militant form of working class political struggle at the turn of the decade, coupled with the pervasive weakness of the British economic base, have made it impossible, for a time, to manage the crisis politically without an escalation of the use and forms of repressive state power. (Hall 1978:284, 304)

It is within this broad context that an analysis of the *Cole-Gittens Report* must be situated. Within this discourse, a legitimacy crisis is created through which the citizen-reader has direct access to the structures of argument that are open to the state within a formal democratic framework. This book highlights the forms, the transmission, the manipulation of elements, and the relations realized in "official" discourses. Based on selective readings, a number of questions need to be raised. For example, (1) How exactly do members or institutions of dominant white groups talk and write about ethnic or racial minorities? (2) What do such structures and strategies of discourse reveal about underlying ethnic or racial prejudices, ideologies, or other social cognitions about minorities? (3) What are the social, political, and cultural contexts and functions of such discourse about minorities? In particular, what role does this discourse play in the development, reinforcement, legitimization, and hence reproduction of White group dominance?

Although these questions focus on "texts" and their cognitive and sociocultural "contexts," the issues raised in this book require a multidisciplinary approach. No single theory or intellectual positioning is capable of comprehending the phenomenal complexity of even a moment of experience, assuming even the possibility of locating and isolating that elusive moment. It is within this problematic limitation that the above questions are asked and a critical discourse analysis is employed in this inquiry. One of the attractions of discourse analysis is that it is able to integrate such a multidisciplinary study of ethnic or racial prejudice, discrimination, and racism. It allows one to make explicit the inferences about social cognitions of majority group members about minorities from the properties of their "text" and "talk."

The methods of political science and the study of law are largely based on discourse. For example, the deliberation of state in decision-making, parliamentary debates, laws, regulations, etc., and also with respect to racial or ethnic concerns is replete with discursive problems. Detailed study of these many forms of political discourse reveals underlying sociopolitical and, in particular, ethnic or racial attitudes of politicians, the strategies of agenda setting, and the manufacture of the ethnic consensus, among many other processes of the politics of ethnic affairs, policing, crime, and immigration, to name a few.

In sociology (ethnography), discourse analysis plays a primary role in accounting for the structures of everyday interaction as illustrated in conversations in culturally variable sociocultural contexts. Thus, the majority of group speakers, or more generally people in Western societies, may engage in the local production and reproduction of White, Western group dominance (James 1996) in communicating stereotypes (Baker 1994) and, more generally, in the reproduction of social, cultural, or political hegemony (James 1996). Such studies are not limited to the micro level of everyday interaction in sociocultural contexts, but also involve macro notions such as groups, social formations (Omi and Winant 1987), or institutions (James 1996), and especially the mass media.

Discourse then, plays a central role not only in "text" studies of the humanities, but also in the social sciences, and virtually all dimensions of the study of prejudice, discrimination, and racism (Baker 1992). Ethnic and racial inequality in social, political, and cultural domains are variously expressed, described, planned, legislated, regulated, executed, legitimated, and opposed in countless genres of communicative and discursive events. Such communication and discourse is not mere "talk and text" of marginal relevance. On the contrary, it is at the heart of the polity, society, and culture in their mechanisms of continuity and reproduction, including those of racism. Lacan once said "There is no knowledge without discourse" (Lemaire 1977:vii). Discourse analysis has displaced epistemology as a form of knowing according to Burton and Carlen (1979:15). This displacement has proceeded in a precarious manner, contradictorily and non-linear in fashion. Traditional discourses have produced one or more savants, including anthropologist Claude Levi-Strauss, psychoanalyst Jacques Lacan, and Marxists Louis Althusser and Michel Foucault, to name a few. Frequent allusions to the works of Foucault (1972, 1974, 1977, 1978) are used, not to provide a direction about how discourse analysis should proceed, but because, in reading this discourse and in the absence of the analyst, it becomes necessary to work within the analytic space, which has made discursive knowledge possible.

What is Discourse Analysis?

Harris provides a useful explanation as to what discourse analysis means. He explains:

> Discourse analysis is a method of seeking in any connected discrete linear material, whether language or language-like, which contains more than one elementary sentence, some global structure characterizing the whole discourse (the linear material) or a large section of it. The structure is a pattern of occurrence (i.e. a recurrence) of segments of a discourse relative to each other. (Burton and Carlen 1979:16)

Discourse is rooted in desire; a desire to communicate with others. Accordingly, one turns to language, as constituted by both the knowing subjects of the discourse—the speaker and the addressee. Referring specifically to the *Cole-Gittens Report*, the Black communities, and, through them, the possible objects of that discourse, consider now, the following quotation from Lacan (1975:61): "The form alone in which language is expressed defines subjectivity. Language says, 'you will go such and such a way, and when you see such and such, you will turn off in such and such a direction.'" In other words, it refers itself to the discourse of the other. But what is taking place within this discourse (inquiry) between the *Cole-Gittens Report* and the Black communities is contradictory, insofar as it is a discourse with, and of, the Black communities directed at capturing future conventionality/forms via an introduction into the present state of the relationship between the Black communities and Blacks in the Ontario Criminal Justice System. It should be pointed out here that this was the first inquiry to utilize anti-Black racism and racial minority women as focal points of analysis of systemic racism. This departs from other inquiries which employ a race-relations approach. Such a departure constitutes a considerable movement toward the question of racism. But, there is a particular absence, an absence of class. This absence of class categories within legal and state forms gives the appearance of administrative neutrality. Those who are in the Ontario Criminal Justice System are "underprivileged" as a result of their own starting point and the individual attitudes they meet in various aspects of life. Thus, the *Cole-Gittens Report* yearns for more Black people to be "middle class" and to have a "real stake" in the community, yet it cannot or does not offer a class theory of racial oppression or a structural theory of any kind.

Interestingly, Foucault (1972) stresses that "unlike linguistic analysis, discourse analysis attempts continually to be non-normative, to deny privilege to conventionality" (p.60). Caution should be exercised here, insofar as such a departure does not change the relationships between institutional and discursive sites of authority or does not neutralize discursive relations. Thus, the language of anti-racism facilitates a new space and simultaneously individualizes the discourse into one that is positivistic in nature. Positivistic discourse, as Burton and Carlen (1979) remind us, "tried normatively to close an artificially reified 'gap' between reality and language, discourse analysis, and is committed to permanent obstruction of such closure" (p.17). The aim here is always to specify particular relationships and conjunctures rather than to erase them by an appeal of or to an ideal order—law and order.

State Apparatus and Official Discourse

Official discourses on law and order are products of the articulation of knowledge as power relations. Like all established discourses, they are signifying practices that demonstrate the effect of ideology on language. An effect is inscribed within a meticulous modality of power. State discourse power is realized in the materialized practices, which Althusser (1971) calls state apparatuses. This text concerns itself with the ideological discursive mechanisms of state legal apparatuses. The concept of ideological state apparatus suggests the association of knowledge and power relations. The objective of such practices, functions via its attempts—successful and unsuccessful and always unfinished—to repair the fractured image of the state's repressive and ideological apparatuses.

It is a well-documented and well-established practice to hold and report on official inquiries into law and order problems. The question of police-community relations tends to capture the public's imagination only during periods of crises. The Younge street "disorder" in the aftermath of the Rodney King trial in Los Angeles in 1992 serves as an example. Events such as these attract disproportionate measures of public attention via the mass media but also serve to highlight the underlying tensions that characterize everyday relations between the police and radicalized minorities in many less-publicized contexts.

Chapter 10 of the *Cole-Gittens Report* deals specifically with community policing. This is not to suggest that the remaining chapters are not important. Stenson (1993) provides a useful distinction that gives some clarity to the concept of community policing. As he points out, "in policy discourse 'community' usually denotes the desire to foster close human links within troubled and fragmented populations, within alienating and fragmented bureaucracies and between bureaucratic agencies of collective security and external social groups" (p. 375). It is clear, then, that the effort to produce a fixed meaning for such a fluid discursive and practical construction is a project of dubious value. The same applies to policing. Policing, whether or not married to the notion of community, functions, for example, as a benign social service (Stenson 1993); as order maintenance (Reiner 1985); as an oppressive force (Scraton 1985); or as part of the attempt to extend the net of surveillance (Taylor 1980) and disciplinary social control into every corner of life (Cohen 1985). Of central concern here is the relationship between the police and some notion of public interest or a public sphere, normally expressed in the form of the state. This may be understood as a benevolent reflection of the general will or an instrument of domination by the powerful. On the contrary, it is argued that "though the police are involved in ruling and may be involved in repressive practices, it cannot be assumed that they are simply a component of a centrally organized or functioning Leviathan, operating according to essential principles" (Rose and Miller 1992:174).

However, the principal means through which social structures are constituted are language and discursive practices that make conceptual distinctions through the play of differences. For one to be in the Ontario Criminal Justice System, a precondition must exist. One must either be a suspect, an accused, or one convicted of an offence. A central issue here is the role of discursive practices. The use of particular ways of talking, as in Cohen's (1985) "control talk," Manning's (1988) "organizational talk," and Thomas's (1988) "law talk," reflects and constitutes what Henry and Milovanovic (1991) call "narratives that provide the continuity to reproduce social structures of crime and its control, in time and space." Henry (1988) explains:

> When state agencies seek to control economic relations that fall outside national tax accounting, they attach derogatory labels to such activity and attribute to it motives carrying negative connotation. Terms such as the "black," "hidden," "underground," shadow," "secret," "subterranean," "submerged" economy are used to suggest that the economic relations of those working "off-the-books" are perpetrated by nefarious creatures of the night who are interested unilaterally in pecuniary rewards incommensurate with effort, who are dishonest, and who cannot be trusted. (P.30)

This, therefore, sets the stage for the control process throughout the criminal justice system.

There is much to cheer about in the *Cole-Gittens Report*. But the apparent attractiveness of this report is based on conservative and theoretically questionable premises. A total of nine major recommendations designed to improve the governance and delivery of community policing were made. The formulation of the basic problem appears on page 336:

> Community policing aims to transform relationships between the police and the community....Many of the challenges facing traditional policing are also found in a community policing system. Among the most important challenges is to respond effectively to public concerns about systemic racism in policing services.

The policing problem, which is part of the objective of this chapter, is a problem *with* the police. Before attempting an answer to the policing problem, consider the "Initiatives of the Canadian Association of Chiefs of Police, such as the development guidelines for community policing in diverse neighborhoods" (*Cole-Gittens Report* 337). Here, we have an indication of understanding the characteristics of neighborhoods. In other words, the social problems will set the standards for successful policing. Now it is safe to claim that standards are variable to hint that police failure by one set of standards in one neighborhood may be understandable by others as a success in a different neighborhood. As Van Dijk (1993) explains, this communication sets the stage for social action and social relations. This speech/talk or expression (Van Dijk 1993) signals various social meanings and categories of social interactions. At the interactional level itself, "respect" and "equality" are signaled as formal modes of implementing community policing. Since "respect" and "equality" markers are mutual here, social power relations seem to be equal in community policing. These initiatives of the Canadian Association of Chiefs of Police also signal social and political dominance in the *Cole-Gittens Report*.

At another level of social relations, that is, relative to the social situation and events discussed, there is no question of formal equality. As noted earlier, the commission is one part of the complex system of criminal justice. In summarizing the "initiatives," the *Cole-Gittens Report* gives considerable weight to the balance of power in terms of what constitutes a social problem to the Canadian Association of Chiefs of Police. This reference allows, or creates, a space for conservative elites, who may otherwise be less interested in community policing, but may still take part in the struggle between racism and antiracism, between "un-Canadian behavior" (p.338) and the values of multiculturalism.

It is clear that the Association, as members of the police institution, speak not only as members of that Association, but with several other social identities, such as White and male. This position institutionally entitles the members to put the policing position on the community policing agenda. Obviously, it is not only their role as Commissioners that influences the structures and strategies of their speech/talk or position, but their identity as members of the White dominant group. However, two paragraphs later, there is a complete reversal:

These preliminary findings led us to focus on strategies for building confidence in community policing among Black and other racialized communities. To develop these strategies, we investigated perceptions of racial inequality in policing, practices that contribute to such perceptions and existing responses to community concerns about systemic racism. (*Cole-Gittens Report*:338-9)

Thus, the problem *with* policing has now become a problem *for* policing. Throughout the chapter, the *Cole-Gittens Report* retains this latter stance. There is a crucial central ambiguity in the argument for "building confidence in community policing among black and other racialized communities" (p.339). It is unclear (perhaps deliberately) whether the "community" is being proposed as a means to an end. That is, the "community" as a new resource for tackling the problems of crime or racism, or, alternatively, the creation of better "community" feeling is the end result, which is being pursued with a concern about crime merely the means of achieving this end. Most of the literature on community policing and community involvement, including the *Cole-Gittens Report,* stresses and cherishes both goals and suggests they are, in some sense, inseparable. But, as we shall see, this failure towards a commitment by the *Cole-Gittens Report,* presents the very difficult choices that have to be made in organizing strategies for increasing community involvement. It conceals the political gulf between those on "the right" who wish to draw on public support to help the forces of law and order; and those on "the left" who seek rather to empower the disenfranchised so that they can confront existing institutions and hierarchies. Above all, the scenario begs the following important questions: If there is already some identifiable sense of community to draw upon, what is its nature and potential? Where community spirit and "confidence" are lacking, what sort of community will be generated by a focus on crime and racism? In general terms, the aim is to make policing more effective by securing the cooperation of the Black communities. On this premise then, the Black and other racialized communities pose problems for the police. This was made clear by a footnote in the *Cole-Gittens Report*:

> Community policing, as it is being discussed across North America, by no means dismisses law enforcement as an important police

function. Rather, it views other methods of problem-solving as more appropriate in the vast majority of cases, and sees these other methods as contributing to more effective law enforcement when the need arises. Police officers are expected to promote communication among those in the community who have conflicting interests and views. Skills gained and relationships developed by police officers through peacekeeping in the community help them deal with more serious problems that require criminal justice processing. (P.336)

The problem here is that racism is theoretically located outside the police agency and its individually agents, who are relatively "blameless victims" of displaced and misplaced aggression among or within communities. Caution should be exercised here, insofar as permitting this theorizing to exonerate the police institution from responsibility for the antagonism felt by Black communities. Furthermore, the society at large, which is apparently responsible, is not given any form; it remains vague and undefined. So, the racism to which Black people are subjected cannot be explained. The problem, therefore, has been displaced onto a shapeless social order in which everyone and no one is responsible. The specific practice of police officers is relatively unimportant. However, the commission found that "many White people share the perceptions of racial inequality in policing; and widespread perceptions of police discrimination are a potential significant obstacle to successful community policing" (p.341).

Then, what about the problems the police pose for the Black and other racialized communities, in particular using their discretion to stop people in cars and on foot? Under "Frequency of Reported Stops" the *Cole-Gittens Report* found that in order to build confidence in the community, the police must find ways to demonstrate that differential stopping of people because of race alone, or in combination with other discriminatory factors, is unacceptable (p.358). This situation is created by the nature of the police services and basic policing methods in law enforcement. Therefore, the following recommendations were made:

> To achieve these goals community policing requires practical guidelines for the exercise of police discretion, training to enable officers to avoid differentials in the exercise of their discretion, and monitor-

ing of police practices. For increased effectiveness, popular education and outreach programs should inform community members of their rights and shared responsibility for community security, as well as the legitimate boundaries of police action. (P.358)

This time, the problem is located not in society at large, but in the subjectivities or false perceptions of Black people, which the society at large causes. Again, this exonerates the police. The conclusion of this style of theorizing is flawed.

There is evidence that Black perceptions of discriminatory treatment by the police are based on real experience and that harassment is probably routine, widespread, and normal. There is further evidence of a methodological problem because the lawfulness of police activities is not an instructive criterion to use when identifying the street harassment of Black or other "marginal" groups. Harassment is lawful, although specifically selecting an ethnic group or race to harass is not. Certainly a search for illegal behavior by police officers will lead to the view that the problem is a minority one. The question, then, is where does the "misperception" lie? It is inherent in the black and other radicalized communities according to the *Cole-Gittens Report*.

Thus, the misperception theory is made possible by displacement theory. This view enables the *Cole-Gittens Report* to assert that the police services and local community organizations, together, should "develop guidelines for the exercise of police discretion to stop and question people, with the goal of eliminating differential treatment of Black and other racialized people" (p.429). As mentioned before, the problem *with* policing is one that individualizes the problem and therefore individualizes the possible solutions. It leads to an emphasis on improved recruitment, improved training, and improved procedures for dealing with the deviant minority and with occasional lapses.

Clearly, all of the above are reasonable suggestions, but the first and last are concerned solely with the quality of individuals and certainly will have no impact on institutional police structural practice that are deeply embedded within society. Improved training will not help if the institution of policing denies that racism is embedded in police discourse. Further, the rotten apple theory strikes a balance between good and bad policing.

However, there is also a danger that exists within this theory. As Cain and Sadigh (1982) argue, "the ideology of police discretion/police professionalism...makes the rotten apple theory and its concomitant individual focused remedies necessary. It is not generated by the rotten apple theory: rather the ideology of discretion/professionalism is consistent with it, and renders structural solutions unthinkable as well as inappropriate" (p.97).

Again, the *Cole-Gittens Report* has thus confirmed the existence of prejudice or racism among the police and within community policing, as have many other reports, but this one has concentrated on racism and prejudice among beat officers. Indeed, does this express a degree of confidence in claiming that racist attitudes would not be sited in the higher ranks of the police? Successive shifts in political language about race since the 1970s have delved into the issue of policing and "Black crime" as a central theme. Whether in terms of specific concerns about robberies, street crimes, or with the question of urban unrest, the interplay between images of race and crime has remained an important symbol in the language of community policing. Discourse about the "Black crime" issue has also been over determined by the phenomena of civil disorder. The latter helps to explain the increasingly politicized nature of the state response to the two issues. The ideological construction of the involvement of young Blacks in armed robberies and other forms of street crimes provide the basis for the development of strategies of control aimed at keeping young blacks off the streets and keeping the police in control of particular areas identified in popular and official discourses as "crime-prone" or potential "trouble spots"

It is at this point I part company with Foucault (1977), in particular the metanarrative in *Discipline and Punish*. His work offers a body politics absent of race. It offers state punishment and prosecution that is considered by some postmodernists to be a master narrative competent in its critique of contemporary state policing. Yet, disturbingly, this work particularly contributes to the erasure of racist violence. As James (1996) argues,

> [Foucault] text illustrates how easy it is to erase the specificity of the body and violence while centering discourse on them. Losing sight of the violence practiced by and in the name of the sovereign, who was manifested as part of a dominant race, Foucault universal-

izes the body of the white, propertied male…it depicts the body with no specificity tied to racialized or sexualized punishment. The resulting veneer of bourgeois respectability painted over state repression elides racist violence against black and brown and red bodies. (P.25)

The attempt here is to show that race is of major importance in any discourse in a racialized society. Race signifies the "criminal" not only by his or her act but also by his or her appearance.[7] Mapping the political terrain is an imprecise craft. However, boundaries are continuously redrawn through political conflict, compromise, and resignation. Losing one's bearings becomes commonplace when following altered maps with abstractions about policing and policed bodies. To romanticize or falsify the disciplined body, one need only present it as unstructured by race, sex, and class. In contrast, rejecting the illusion of an individual in a casteless society composed of raceless and genderless bodies, one may confront racist and sexist violence in state practices and theories that mask such violence.

The critical and practical relevance of this book is that in situations where tolerance, equal rights, and the rule of law are officially respected, discourse may subtly signal contradictions. An anti-racism approach does not set out to discover the factors which have contributed to a Black person's predicament but to explain how, in any given situation, racism has created that predicament. It starts with an answer and works its way back towards the question. Anti-racism, by adopting an idealistic paradigm and a simplistic view of racial oppression, has steered us away from those questions, which could enable us to develop an adequate explanation of the phenomenon. An adequate explanation must account for the interplay of race, class, gender, social structure, culture and biography, and the ways in which they shape the chances and choices available to groups and individuals in similar structural locations. Even moderate feelings of superiority, stereotypes, prejudice, and relations of social inequality defining "modern" racism may be involuntarily presupposed, expressed, or signaled in text and talk. It is a critical discourse analysis that may literally reveal processes of racism that otherwise would be difficult to establish, or that the majority would formally deny. In this respect, a critical discourse analysis may yield an instrument or confirmation of counter ideologies, which, in turn, support dissent and counter power.

Interestingly, researchers continue to produce different findings, and that differences are found within the same study, means that any discrimination is attributed to the exercise of discretion. Racial prejudice on the part of individual police officers is put forward as the explanation and the target of intervention, rather than more structural factors. However, the social structural factors in the sense of social position of Blacks or other minority ethnic groups, or the way that criminal justice itself is structured, is a seemingly absent consideration. As Harris reminds us, "if a prima facie case of discrimination can be sustained against the criminal justice system, it is a case against individual practitioners which has been conceded, not against the system as a whole, still less the social role of criminal justice and criminalization" (Harris 1992:110).

Notes

1. The Critical Race Theory movement consists of scholars who are challenging not only traditional legal paradigms but also critical movements that have developed and evolved in essentialist ways. For the most part, the focus has been on a critique of domestic initiatives, laws, and normative mythologies. Patricia Williams (1987) examines racism as a crime. She considers how "the rhetoric of increased privatization in response to racial issues, functions as the rationalizing agent of public unaccountability, and ultimately, irresponsibility" See "Spirit-Murdering the Messenger: The Discourse of Finger-pointing as the Law's Response to Racism." *The University of Miami Law Review* 42(147):127-157.

2. Here, Normandeau and Leighton (1990), and *The Report of the Commission on systemic Racism in the Ontario Criminal Justice System* (1995) serve as useful examples both at the Federal and Provincial levels of government.

3. This is a matter of importance for Caribbean peoples since criminalization has been a key mechanism of social control. *Bill C-55, An Act to Amend the Criminal Code (High Risk Offenders) and sections A70 (5), (6) & 77(3.01)* of the *Immigration Act* illustrate ways in which criminalization is further expanded through citizen and non-citizen.

4. This involves the central government seeking to act upon crime not in a direct fashion through state agencies (police, courts, prisons, social work, etc.) but, instead, acting indirectly, seeking to activate action on the part of non-state agencies and organizations. Its key phrases are terms such as "partnership," "inter-agency co-operation," "the multi-agency approach," "activating communities" creating "active citizens," and "help for self-help." See David Garland (1996).

5. This *Cole-Gittens* Report was undertaken as a result of an initial investigation into racism in Ontario conducted by Stephen Lewis at the request of Premier Bob Rae. The *Stephen Lewis Report on Race Relations in Ontario*, June 1992, described the prevalence of racism in the Ontario criminal justice system, education, and employment. Further, the Report found that the primary focus of racism was in the Black *communities*.

6. Here I am referring to Antonio Gramsci's notion about

> the organizing principle of a society in which one class rules over others not just through force but by maintaining the allegiance of the mass of the population. This allegiance is obtained both through reforms and compromises in which the interests of different groups are taken into account, and also through influencing the way people think.... This enriching of the meaning of hegemony is related to the increasing complexity of modern society in which the terrain of politics has changed fundamentally. In the age of mass organizations such as political parties and pressure groups, when the expansion of the suffrage requires any state, however restricted democratic liberties are, to attempt to maintain the consent of the governed—and with the development of the educational and cultural level of the population, its ideas, practices and institutions—the area of the state action expands and the private spheres of society are increasingly intertwined. In this context the very meaning of political leadership or dominance has changed, as rulers must claim to be ruling in the interests of the ruled in order to stay in power. Increasingly the demands and needs of society have come to be considered the responsibility of governments, when once they

might have been relegated to the private sphere defined as outside politics. Ideas, culture and how people view themselves and their relationships with others and with institutions are of central importance for how a society is ruled and is organized, and underpin the nature of power—who has it and in what forms. Thus, as the very nature of politics has changed, the meaning of hegemony, as leadership, dominance or influence, has in turn evolved. It now also implies intellectual and moral leadership and relates to the function of systems of ideas or ideology in the maintenance or the challenge to the structure of a particular society. It is consequently instrumental not only in the continuance of the status quo but also in the manner in which a society is transformed. (Anne S. Sassoon, *The Blackwell Dictionary of Twentieth-Century Social Thought*, 1994, pp.225-226. Edited by William Outhwaite and Tom Bottomore. Oxford: Blackwell)

7. A 1994 television documentary by Michael Moore, producer of *Roger and Me* illustrated how in the American mind criminality is constructed as a racial marker. The producer videotaped an African American middle-class man attempting to hail a taxicab while a block behind him a European American man, an ex-convict who had served lengthy jail sentences for violent crimes, also tried to hail the same cabs. Overwhelmingly, the taxi drivers bypassed the Black man to pick up the White man. In this racialized society, White convicts (and ex-convicts) exhibit a higher social status than Black non-criminals and criminals. Whiteness exculpates and signifies the "normal," just as Blackness implicates and marks deviance.

Chapter 2

Personal and Up Close: Young Black Men Speak on Crime, Race, and Community Policing

A Methodological Consideration

Forests of paper, volumes of press ink, and millions of research dollars have been devoted to the plight of the young Black male in North American society.[1] He has been scrutinized, objectified, and memorialized; so much so that for most of us "young Black male" is synonymous with criminality. The negative criminal label is applied as though one characterization could accurately define such a vast group. Although some people have suggested that a generation of Black men has been lost to the criminal justice system by way of conviction or under some order of the court, in reality, the criminal stereotype describes only a fraction of the entire group. A commonly quoted statistic is that 1 in 3 young Black men are under the jurisdiction of the criminal justice system (*Cole-Gittens Report* 1994 and Mauer and Huling 1995). What often goes unacknowledged is that if 33.3 percent are in the justice system as accused or convicted persons, then 66.7 percent are not. Specifically, very little media attention is focused on those who are enrolled in college or university.

Law-abiding young Black men are frequently overlooked as a resource for analyzing crime and justice issues. To tap this source, focus groups were conducted with young Black male university and college students. While an increasing number of books explore the breadth of black male experience, very little research presents their specific viewpoints on crime, race, and society.

All Black men enrolled in "Crime and Delinquency and Race" and "Minority and the Legal Order" courses at a Canadian university were invited. Black men from a college, who were enrolled in a *Law Enforcement Certificate Program*, also participated. Fifteen men in all agreed to participate, ten from university and five from college. The focus group sessions were held at the university in June 1994. Two groups, one with eight participants, the other with seven, each met for one-and-one-half hours. Each participant received five dollars for transportation. The author and a colleague, a Black female graduate student in sociology, led the focus group discussions. Audiotapes were used. All names used for focus group participants are aliases.

Focus Groups

Participants were first asked to share their reactions to receiving an invitation to participate in a focus group about young Black men. Several young men commented that they were "excited" about the invitation to participate in this research. One student said it is "rare" for the perspectives of young Black men to be solicited. Each participant had their own interpretation of "race, crime and community policing" when asked the question "What does race, crime and community policing mean?" Most respondents replied that it is synonymous with "Blacks, crime and community policing." Jack, a 22-year-old from Scarborough, Ontario, with a family history in law enforcement, said the terms are used as a way to "reinforce stereotypes… that certain crimes are committed by Blacks and for the police to be heavy-handed". Noel, a 24-year-old, North York native said, "When I hear those phrases…I automatically think they're associating Blacks [with] illegal activities and an excuse for the police to spy on us." Robert, age 22, born in Guyana, commented, "It makes me think of something my mother always says: 'They got another Black child on television'. With crime, the media emphasis is always on Blacks." Bob, 25, from Lawrence Heights, North York, echoed Robert's sentiments and said, "Many people don't realize that the media is the only way most Whites [see Blacks]."

Reflecting on the Criminal Justice System

More than any other race and gender group, Black men have the greatest probability of contact with the criminal justice system, as both victims and offenders. The young men responded passionately when asked to comment on the criminal justice system. "Racist" and "biased" were the terms invoked most frequently. Clinton, 27, from Mississauga, Ontario, said, "It [the criminal justice system] is just a microcosm of what the larger

Canadian society is...[It] has been more or less an instrument to maintain the racism within the country."

Roy, a 21-year-old born in Jamaica and living in Canada since the age of 5, said, "My mother always told me that the criminal justice system is like slavery. When slavery was abolished, White people tried to find new ways to bring Black people down." Charles, 25, born in Canada with parents from Jamaica and living in Vaughan, Ontario, referred to the criminal justice system as "a sinking ship," and said, "We know certain [policies and strategies] do not work, yet we continue to use them." According to Johnny, when the rights of Blacks are violated through community policing without recourse, it is evidence of racism in the criminal justice system. He compared the constitutional rights of young Black men to an obstacle course, "[It is] almost like being on one side of a [fifty-foot] wall and you need to get to the other side without being fifty-feet high. What can you do? You try to find a way around it but there doesn't seem to be one."

Hearing Stories, Creating Boundaries

Given that young Black men have the greatest probability of falling victim to other young Black men (Wilson 1990), it would be reasonable to expect that they would be fearful of one another. The focus group participants said that whether they would be fearful of other young Black men would depend upon the situation. Roy made the following statement: "I'm not fearful if I see a Black brother on the street. Really, I'd be more afraid of a White guy than I would a Black guy, simply because I [have] some relationship to the Black guy. I have no problems with one Black guy. But if I see a large group of them with joints [marijuana] in their mouths, wearing gold, I'd be afraid." Several young men in the focus group echoed Roy's sentiments. Some noted that if they were traveling alone, the presence of a group of young men, of any race, might be cause for concern. Factors contributing to fear included the behavior of the individual or group, the time of day, physical size and conduct of the approaching person, and how the person is dressed. Overall, the focus group appeared to be fairly discriminating in assessing what constitutes a threat. It may be, I would add, that because these young men are frequently targets of unfair negative stereotypes, they tried not to unfairly stereotype other young Black men.

The participants were asked how old they were when they became aware of the negative public perception of young Black men. Gary, 24, from Mississauga, was one of the men who said he was aware of the stereotype by age 9. Born and raised in a small, predominantly White town near London, Ontario, he recounted a frequent childhood occurrence: "At the playground in the school yard everything is cool [until] an argument comes up. Then, all of the sudden, you're the nigger." Several others said it was comments made by Whites that first made them aware of the negative image associated with Black men.

In response to the question, "How does the stereotype affect you?" most of the men commented that they have grown weary of the negative labels. Others expressed anger. Shawn, from Scarborough said, "It becomes annoying.... I don't want to hear about another young Black male and the police." Eon, a 25-year-old from Etobicoke, said he is angry at Whites and Blacks: "I'm upset about the negative portrayal of Black men on TV. I'm upset at Whites and Blacks. I'm upset at the media because they [say that] Blacks commit most of the crime. That's not true...but I am upset with Blacks because we perpetuate a lot of the stereotypes ourselves. This gives the police a good reason to be trigger-happy and then defend their action[s], by saying that there are fighting crimes and the public believes this. Look at all the foolish movies we've made, like *Boys in the Hood* and *Menace 11 Society*."

The participants also discussed their perceptions of law enforcement. Most of the men had heard negative stories about the police. Ten of the fifteen participants came to the conclusion that the police were hired to enforce the law against Blacks. Winston commented, "I was always told the police are only good people. Really, it is made up of people who lie, people who cheat, people from all walks of life. It is made up of guys off the street." In school, Winston learned about "Officer Friendly" and was taught that the police are the "good guys." The picture of the pleasant, helpful, resourceful police officer, however, contradicted his direct experience: "It did not take long to realize that the police were not our friends. When I was a child, they would come to our block and tell us not to play basketball in the in the street. They would tell us to 'Go to a park'. Well, we didn't have a park to go to and the White kids played hockey in the street, so that was harassment to us."

Fear and Respect

None of the young men directly acknowledged being fearful of the police. Nevertheless, fear is the word that best describes their motivation for avoiding law enforcement officers. The incongruous result is that the young Black men interviewed are more fearful of the police than of other young Black men. The result is incongruous because they represent the race and gender group with the highest rate of criminal offending, victimization, and being shot by police (*Cole-Gittens Report* 1994). Black men aged 16 to 25, are the most likely to need police assistance. However, Black fear of police, and the informal code of silence that it may provoke, has implications for police relations with Black communities.

Another aspect of the fear dynamic is how others perceive Black men. The participants were asked, "In public, how do Whites respond to you?" Most respondents said that Whites are afraid of them. Some cited the never-ending insults that go along with being perceived as dangerous, the difficulty getting directions from a stranger, hearing the click of automatic car locks as you walk by, being stopped by police, being followed in a store, and interrogated about crime. The actions often taken by Whites or Asians to protect themselves or their property from Blacks are either aggressive moves "towards" Blacks (e.g., watchful eye of a shopkeeper) or aggressive moves "away" from Blacks (e.g., moving away from them on the street).

Strangely enough, the young Black men describe White fear in ways that could be easily confused with White respect. Roy, who is 5' 8" and stocky, expressed this: "Out in public I will see a White guy coming towards me. He may be 6' 5" and over two hundred pounds and all of the sudden he will step aside and let me pass by. I am shorter and less stocky, but [the White guy] just steps aside and [lets] me pass, no matter what I'm wearing." This kind of reaction from Whites could be interpreted as either respect or fear. Other participants recounted similar experiences of deference from White men and women. At the same time that Black men are cast off as being criminals, they are also celebrated as being hyper-masculine. They are idolized as cool, hip, and sexually gifted. These contradictory reactions may encourage some young Black men to capitalize on their macho image. Several of the young men said they believe that some Black men turn to crime and deviance as a self-fulfilling prophecy. That is, many engage in crime because people expect them to. However, none of the participants named a specific crime. This suggests that deviance and crime were used interchangeably. Some of the participants expressed concern that crime by Black men is glorified and idolized by Blacks and Whites alike.

What lessons should we take from the comments of these young Black men? While they have high expectations for themselves individually, they feel encumbered by the negative labels attached to being young, Black, and male. A few pondered aloud how high they would be "allowed" to climb on the economic ladder. Their reflections and experiences with race, from their feelings of invisibility and fear to their feelings of frustration and anger, painfully illustrate the costs of negative stereotyping. The interviews provide us with an understanding of what it is like to be an involuntary, life-long representation of deviance or a "dangerous supplement" (a notion that will be developed in the chapters that follow). The focus of the media on the small percentage of young Black men who are criminals has exacted a burdensome toll from the majority who are law abiding. The non-criminal majority is an untapped resource. It may be that the young Black men who do not fit the criminal image can help us understand those who do.

Discussion

Criminologist Daniel Georges-Abeyie (1990) asks, "Does the focus of criminal justice analysis on the formal, easily observed decision-making process obscure or even misdirect attention from the most significant contemporary form of racism within the criminal justice system?" (p.12). He further states "that an examination of the formal stages is insufficient to determine the prevalence of racial bias" (p.32). The act of stopping motorists, which constitutes an informal stage, determines in large measure who will be arrested and thus who will enter the criminal justice system. Accordingly, the above encounters, which are not subject to official measure, must be included in this book. These data help reveal the nature of support for community policing and, equally, how consent gets constructed.

The direct and indirect experiences that Blacks have with the police affect their perception that the balance of criminal justice is tipped against them. Many people would argue that it is unfair to blame the police for being suspicious of Black men. After all, [let's assume] Black men are disproportionately engaged in crime. Then it is reasonable, that the police disproportionately suspect them of criminal activity. Again, (let's assume) Black men do commit street crimes at high rates—rates far exceeding their percentage in the Canadian population—the important question, in spite of our assumption, is "are Black men stopped and questioned by the police at a rate that greatly exceeds their rate of street crime"? If so, the number of police stops cannot be legally justified.

The available research suggests that Black men are stopped and questioned at a rate much higher than the level of their involvement in crime. One way to determine this is to compare the rate of police stops for Black men with the rate of Black men who are involved in criminal activity. For example, [assuming] that one-third of all young Black men are involved in crime, we would predict that one-third of them would be subject to police stops. The problem, here, is that it would be difficult to gauge what percentage of Black men are "involved" in criminal activity. The incidence of non-reported criminal activity and criminal activity evading arrest and the "Dark Figure" of crime pose a problem. This is an area that is in need of much research. The available evidence indicates that the police stop more than one-third of all young Black men. As yet, no national data have been collected on the incidence and prevalence of police harassment and abuse.

The "Why"

So-called "objective" research does not exist. Research is influenced by the researchers' race, gender, class, their values and principles, and their academic and social environment. With this in mind, the information for this book was gathered over a six-year period, which in effect allowed participants to become *ipso facto* researchers. To meet this goal required a method of research that involved producing knowledge differently, one that would ensure that the needs and interests of the Black communities would drive the process and not just the author's needs and academic interests.

As a result, an approach was devised that validates experiential knowledge, and approach where communal knowledge is valued as opposed to the traditional methods that give primacy to academic, rational knowledge. This is not to dismiss in a rudimentary fashion the use of qualitative and quantitative research that is still based on a narrow understanding of what constitutes and generates knowledge. While experiential knowledge may be flawed in a number of ways, for example, a respondent overgeneralizing from personal experience, it serves to give a voice to those racial events that goes unreported. This method is in need of much further development in the study of exclusion.

The Results

The results of this method may be measured in terms of the changes that are initiated in the lives of individuals and changes within the Black communities. These range from some Black women deciding to go to community meetings to some Black men recognizing that Black women also face the same problems of police harassment and brutality. The knowledge produced by the group was useful knowledge, which directed action within the lives of those involved. The process also led to interactions and discussions between the older generation and youths around issues of sexuality, class, and nationality and its relevance to racism and community policing. On a larger scale, this work prompted more liaising with other communities as well as political activism to ensure adequate services for the communities.

Note

1. It should be noted that police practitioners and academics in Canada have drawn on studies from other countries to inform their policies on policing but there is clearly a need to initiate Canadian research. Unlike Britain, which has the Police Foundation and the Home Office Research and Planning Unit, and the United States, which has the Police Executive Research Forum and the Police Foundation, Canada does not have a national facility for promoting and initiating policing research. Since the April 1993 demise of the *Canadian Police College Journal*, Canada does not have a vehicle for the dissemination of the results of projects conducted in Canada. Although the comparative lack of research can be attributed in part to Canada's smaller population, without an autonomous institute for Canadian police research similar to those in other Western democracies,

police research will remain fragmented and peripheral, and primarily conducted in university departments, the Canadian Police College and under contract to federal and provincial governments, subject to haphazard funding. Theories of policing may be taken from other jurisdictions and applied in Canada, but empirical analysis must be located in Canada. See Seagrave (1997) *Introduction To Policing In Canada.*

Chapter 3

Criminalization: Racializing Crime

The Color of Crime

The media portrayal of Blackness is primarily depicted through images of Black men. As discussed in this chapter, though the depictions are mostly ones of deviance, there are also enough images of Black success to create public confusion in assessing how Blacks fare in American and Canadian society. The contradictory media portrayals of Blackness are critically analyzed and followed by an inquiry into the interpretations of young Black men's public images and how they themselves view the criminal justice system. At a time when young Black men are the focus of much criminal justice attention, it is critical to discuss their assessment of their image and their reality. Lastly, this chapter concludes with a unique look at Blackness through community policing.

There is a lack of awareness of the forces that shape our perception with regard to how we make sense of our daily experiences. In particular, our attitudes toward crime are molded so that we are inattentive to the processes that construct our standpoint. The media play a considerable role in shaping the perceptions of crime within our communities; that is, in a (re)selective reconstruction as to what constitutes a criminal event. Crime as reconstructed in the media promotes the practices of the elite class while criminalizing the exploited classes. Thus, the media content is primarily ideological rather than of entertainment or educational value, which is ostensibly paraded. The mass media are important agents in the reproduction of ideology because of their central role in socialization. Therefore, crime events are selective ideological reconstructions that reproduce, repair, and maintain the inequalities within society.

Although the majority of us have never been the victim of a crime or been charged with a crime, we all have certain images of crime. There are certain perceptions that have been created with regard to class, race, and crime (Baker 1994; Russell 1998). It is both the best and worst of times for representations of Blackness in the media. On television, Blacks are regularly portrayed as lawyers, doctors, nurses, police officers, and best friends. In fact, more Blacks appear in print, radio, and television journals than ever before. At the same time, however, television programs continue to feature updated versions of stereotypes that have existed for centuries. In North America, Blacks have been criminalized as a race. McIntyre (1993) tells us that "European Americans created a social structure for free Whites and enslaved Blacks and viewed free or freed Blacks as unwanted, troublesome and dangerous, inherently criminal...and uncivilized, and relegated us to the lowest social, economic and political class of their society" (p.21). In contemporary society, the media function in a particular way to carry on that false legacy, thus, when the police brutalize or shoot a Black person, the victim is already criminalized. It is not unusual to hear statements such as "they must have being breaking the law." Thus, Blacks are stigmatized as the "dangerous supplement" (Hall et al. 1978; Trotman 1986).

Hall et al., in *Policing the Crisis* (1978), have analyzed the history of the media and popular responses to the mugging issue. The premise of that study was that the construction of Black communities as a social problem was the ideological bedrock on which the Black youth/urban deprivation/street crime model of mugging was constructed. Mugging as a political phenomenon, according to Hall et al. (1978), became associated with Black youth because they were seen as (a) a social group which suffered the most direct impact of the cycle of poverty, unemployment, and social alienation that affects inner city areas; and (b) suffering from the added disadvantage of belonging to a racial group with a "weak" culture and high levels of social problems, such as broken families and lack of achievement in schools. The power of these images, according to this study, derived partly from popular common-sense images about race and the inner cities, but also from the feelings of uncertainty, which were developing within society as a whole about the position of black communities and their role within the dominant institutions (Hall et al. 1978:346-9).

Hall et al. (1978) note, for example, that even in areas where young Blacks were a small minority of the total youth population, the issue of crime on the streets became intimately tied with the category of Black youth. This ideological construction was possible because the dominant concern about the "ghetto areas" focused on the supposed drift of young

Blacks into a life of crime and poverty. Interestingly, in Canada, more specifically Toronto, the discourse for the last twenty years followed similar lines. Toronto has long been regarded as a model among North American cities (Lemon 1984). Caricatured as "Toronto-the-Good," the city that works, its crime rates are well below those of comparably sized U.S. cities (Jackson 1994). However, Toronto's reputation for safety and tolerance has been undermined by a series of events involving police officers and members of the city's Black (West Indian) communities that have led to a serious deterioration in police-community relations. These events, such as policing shootings, have prompted a questioning of Canada's commitment to multiculturalism, as institutionalized in legislative measures such as the *Human Rights Act, The Charter of Rights and Freedoms*, and *The Multiculturalism Act* (Kobayashi 1993).

Toronto's somewhat complacent self-image in terms of race relations is epitomized in one of the city's earliest report on police-community relations, Walter Pitman's Report *Now is not too late*. This is one of the well-known reports often referenced in relation to the early 1970s. Walter Pitman explained the positive image of Toronto:

> Toronto became world-renowned in the early 1970s as the great metropolitan city in North America, which has "made it." In place of racial riots, citizen alienation, traffic turmoil, this city had come through the '60s with a reputation for the preservation of its neighborhoods, the creation of an efficient public system for police, fire and transportation services, and most important of all, that people of all colors, ethnic backgrounds and religious traditions could enjoy its public places in safety. (Pitman 1977:22)

However, from around 1989, this favorable image became harder to support as a series of incidents involving conflicts with the police resulted in the death or permanent injury of several Black people, often in very disputable circumstances. *Maclean's* magazine, reflecting on these events, ran a cover story in January, 1989 entitled "Police Under Fire," referring to the barrage of criticism that had followed police shootings in Toronto and other Canadian cities. The daily press ran a number of similar stories with a variety of controversial illustrations including the *Toronto Sun's* cartoon of a gun-toting Black man cheerfully handcuffed to a White police officer (16 January 1989). During Toronto's civic election in October 1991, tensions resurfaced. The issue of "Race and Crime" again made the headlines (*Toronto Star*, 19 October 1991). In this case, the controversy was about the collection of crime statistics, with the accusations of racism being directed at those who favored the identification of offenders by racial group. Then, following the Rodney King verdict in Los Angeles, a peaceful protest march degenerated into a "rampage" along Yonge Street in downtown Toronto, reported in *Maclean's* under the following headline: "Young Black and Angry: A Toronto Riot Spotlights a Season of Urban Tension" (18 May 1992). This was followed by a series of official reports, including a specially commissioned report on Race Relations in Ontario (Lewis 1992) and an audit of the "race relations" practices of the Metropolitan Toronto Police Services (Andrews 1992).

Police-Community Relations

At this point, a brief overview of police-community relations in Toronto at that time is warranted in order to contextualize the problem. Before community-police relations could come to dominate community relations, a set of associated changes were required to guarantee the new order. The existing community boundaries had to be transformed from an intra-community relations structure to a structure of inter-community relations in support of community policing. Intra-community relationships, embodied in monopolies of all kinds, cut into the control capacity of governmentality. The Black communities had to be made responsive to the needs of the state apparatus, by removing social control over the communities (intra-community control) by reducing the hold of community leaders over the communities. At the same time, state investment in voluntary organizations or Community Boards of Directors had to be redirected toward the creation of an infrastructure of communication that could benefit policing without demanding excessive expenditures from the government. Toronto is now the most ethnically diverse city in Canada, with 42 percent of the country's minorities, up from 9.4 percent in the 1991 census (Statistics Canada, 1996 census). Such diversity is a fairly recent phenomenon. However, Toronto's population remained predominantly of Protestant and British descent

(Lemon 1984). The development of the city's police services clearly reflects this highly sectarian past.

Official attempts to improve police-community relations in Toronto date back to the creation of an Ethnic Relations Unit within the Metropolitan police in the early 1970s. At that time, the Unit was principally concerned with "the urgent need to understand the problems faced by recently arrived Italian immigrants" (Metropolitan Toronto Police, no date, p.1). The Unit then changed its name to the "Inter-Community Relations Unit" and set up a Black Section in 1975, followed by sections for various other minority groups in subsequent years. Added to these separate initiatives, the Police Services Board (formerly the Police Commission) adopted a comprehensive Race Relations Policy in August, 1989, including statements on community relations, employment equity, staff development and training, media relations, and public complaints (Metropolitan Toronto Police Services Board 1989). Ironically, this concurred with a rapid deterioration in relations between the police and the Black community following a series of police shootings.

Much of the public debate about these incidents in Toronto focused on the role of the Black Action Defense Committee (BADC), which formed in 1988 to coordinate protests against the killing of Lester Donaldson. The BADC was given a hostile press reception, in particular Dudley Laws. Mr. Laws was referred to, in *Toronto Life,* as "Toronto's most infamous radical Black activist" (August 1989, p. 30). Mr. Laws described the Toronto police as "the most brutal, murderous force in North America" for which he was immediately sued for defamation by the Metropolitan Toronto Police Association (*Share*, 18 April 1991). However, Art Lymer, then President of the Police Association, called BADC "a small group of extremists that does not represent or benefit the black community" (*Globe and Mail*, 14 January 1989). As was shown, BADC received support from a number of community groups including the Canadian Jewish Congress, the Chinese Canadian National Council, and others (*Now,* May 1990, p. 24-30).

After manslaughter charges were brought against P.C. Deviney in the Lester Donaldson case, Police Association President Art Lymer declared that police morale was at an all-time low. Under such circumstances, Lymer called for the resignation of Attorney General, Ian Scott. He argued that the charges against P.C. Deviney were "a political move" inspired by "a small minority of Black activists" who were not representative of "the vast majority of the black community which is law-abiding and supportive of the force." Some officers, he claimed, had threatened to turn a blind eye to crimes committed by Blacks: "Police officers will be reluctant to act and to arrest Black people and they'll just take over the city and you'll be back to Detroit" (*Globe and Mail*, 17 January 1989). Far from condemning these remarks as undisciplined or unprofessional, the public rallied around the police. This was demonstrated by way of phone-in programs on radio and television of an overwhelmingly supportive public (of the police) and public demonstrations by groups such as Citizen Opposed to Police Slander (COPS) (Jackson 1994). Concerns also mounted as some local politicians supported the BADC's criticisms of the police. Howard McCurdy (the only Black member of the House of Commons at the time) argued that "police forces are unresponsive to the communities they're serving," and that "many people feel that police are beyond the control of the community and act as if they're above the law" (*Globe and Mail*, 9 February 1989). In order to consider a broader range of opinion, we now turn to the evidence submitted to the 1989 *Task Force on Race Relations and Policing*.

On 14 December 1988, following the fatal shooting of Michael Lawson, the Ontario Solicitor General, Joan Smith, appointed Clare Lewis, the civilian head of the Metropolitan Police Public Complaints Commission, to chair a Task Force to investigate police community relations throughout Ontario (Ontario 1989). A variety of methods, including a questionnaire survey of all police forces in Ontario, an invitation to submit written briefs, and a tour of the province to receive oral testimony from concerned groups, resulted in 127 submissions from a wide range of groups and individuals. This documentation offers direct insight into prevailing discourses of "race" that embrace various different "constructions of criminality."

The Task Force drew attention to the grievances expressed by many minority groups and suggested that the police may be failing to honor their pledge of serving all members of the public equally (Ontario 1989). The Report also highlighted complaints of police insensitivity toward visible minority communities and provided further evidence to support the charge of differential policing. Like the Scarman Report on the Brixton "disorders" in Britain (Scarman 1981), the Lewis report did not sustain the charge of institutional racism.

Analysis of Evidence Presented to the Task Force

What follows is an analysis based on a selective reading of the Task Force evidence. However, material has been included from a range of perspectives that highlights a number of different constructions. For example:

> This country...was settled by Europeans/Christians people and we intend that it remains this way (Task Force on Race Relations and Policing, 1989; written brief no. 106)

> The police should not be on trial for the shooting of these young criminals...Every country that let blacks in have lived to regret it...Jamaicans produce poor unfortunate bastards by the thousands...Get your patriots straight. (Written brief no. 32)

The relationship between minorities and the rest of Canadian society was also frequently commented upon, for instance:

> The vocal minorities that have sparked your Task Force are but a small part of the overall population. I fear in order to pacify them you may alienate the police and the majority of people they serve. (Written brief no. 41)

> The question of "cop culture" (Reiner 1985) was often discussed as one of the mechanisms through which police attitudes toward minorities are formed. This contribution from someone who was involved in police-community relations training is typical:

> When this type of training took place...there was a certain amount of reinforcement of unproductive attitudes through conscious or unconscious peer pressure that could be described as ethnocentrism. (Written brief no. 3)

One black parent in submission offered a sensitive analysis of how the press had responded to criticism of the police:

> Each time there is an outcry for justice from the Black community, there is an attempt to discredit those who come to the fore, by labeling them "agitators." The Dudley Lawses of this society have as much right to peaceful protest as the Art Lymers... Dealing with attitudes of the police, it is a commonly held view in the police service that the police should merely enforce the law without concessions to any one section of the community. Fundamentally, this is not an unfair stance, but in reality it is much too simplistic. When any group of people with different aspirations and cultural backgrounds adopt a style of living, apparently at variance with the norm, it is clearly just not good enough for the police to aver that all must be treated alike. It is right that the integrity of the law should be preserved, but the means to achieve this end can be different. (Written brief no. 33).

Apart from these specific examples, which give some insight into the different ways the relationship between "race," crime, and policing are conceived, newspapers and other media provide further insight into popular constructions of criminality. In May 1990, the *Globe and Mail* ran a feature article entitled "What Do Police See Encountering Blacks?" The article included material from an interview with a Black lawyer who argued, "Police are afraid of Blacks." The lawyer explained, "They know intellectually that not all Blacks are crooks, but emotionally, in their gut, it's a different story...Toronto police live almost exclusively in a White world and they Police black neighborhood like an army of occupation."

In 1986, a military metaphor was used when police swept through the predominantly Black neighborhood of Lawrence Heights in Toronto making ten arrests. Local critics of the raid wrote to the *Toronto Star* (30 October 1986) that their community had been "treated as if it was occupied by a foreign army." This was not an isolated incident. Other Black neighborhoods had been similarly targeted including a police raid on the Malvern Christian Assembly Church in January 1989 and a drug sweep of the Jamestown area in north Etobicoke in July 1989. In both of these cases, members of the Black community described events in terms of an "invasion" or "siege" (*Toronto Sun*, 11 January 1989; BADC press release, July 1989).

Additional common constructions include a symbolic opposition between the police and Black people such as that reported in the popular magazine *Chatelaine* (February 1991) under the headline "'White Bias' or 'Black Crime'? Toronto's Race Crisis." Interestingly, the police are themselves treated as a kind of visible minority, closely paralleling the way the Black community is represented. According to this construction, the police are seen as a recognizable (quasi-ethnic) group, with a distinct appearance, a common occupational culture, strong internal loyalty, and organizational coherence. Here, police-community relations are treated in terms of a theory of mutual mistrust that occurs "naturally" between all such groups. This view is clearly represented in the following quotation from a Haitian police officer in Montreal:

> The police force and Montreal's Black communities are both suffering from the same syndrome: people are assuming from the criminal actions of a small minority within the Black community that the whole community is criminal. In the same way...there's an assumption that the racist actions of a small minority within the police force mean the whole force is racist. There's a mutual incomprehension...that no one seems to be correcting. (*Montreal Gazette*, 27 July 1991)

Such constructions are widely supported within professional and social science literature on police-community relations. While there may be a fragment of truth in such common-sense arguments, such arguments fails to appreciate the unequal power relation that structure every interaction between the police and visible minorities, which makes the argument that the two groups suffer from the same syndrome quite untenable.

Crime Statistics

Current debate about race and crime statistics in Toronto goes back to 1988, when Staff Inspector Julian Fantino commented to the press about the level of crime within the Jane-Finch area, a low-income suburb well-known for its relatively high concentration of Black residents. Mr. Fantino maintained that Blacks were overrepresented in certain kinds of crime. Constituting just 6 percent of the neighborhood population, he alleged that Blacks were responsible for 82 percent of robberies and muggings, 55 percent of purse snatching, and 51 percent of drug offenses (Jackson 1994). A similar debate was taking place in Britain at this time, as to the construction of mugging as a specifically "Black crime." While it was widely asserted that "robbery must be classed as a predominantly "Black crime," Hall et al. (1978) had already challenged this construction, arguing that "the mugging crisis was a form of moral panic, orchestrated around a politically and media-inspired crime wave" (Hall et al. 1978:327). Despite the sensitivity of the issue, police officials repeatedly made provocative statements, such as Police Association President Art Lymer's unequivocal remarks on "Black crime." In an interview with *Chatelaine* magazine (February 1991), he argued, "there is a problem with Black crime out there...95% of crack is being distributed by Blacks."

The debate about crime statistics has polarized into those who feel that such figures are needed in order to demonstrate the fairness of the criminal justice system and those who feel the data would lead to false inferences about the nature of racialized crime (*Toronto Star*, 12 and 19 October 1991; *Globe and Mail*, 8 February 1992). However, the issue cannot be divorced from and must include the wider debates over the existence of "institutional racism" within the police services. In Toronto, as elsewhere, there has been an unrelenting reluctance to acknowledge the possibility of police racism even as a reflection of the wider society from which officers are drawn.

Cultural Racism

In addition to the many ways in which racism has been expressed traditionally—overt and covert, direct and indirect, and intentional and unintentional—racism is increasingly expressed through an apparently non-racist code. The result of such "cultural racism," according to Balut (1992), is "there is still much racism but few people who can be unequivocally identified as racist" (p.297). A key contemporary form of cultural racism is the association between certain geographical locales and particular racialized groups (Joseph 1984). The process is exacerbated as many police officers that work in these areas live in all-White neighborhood and have little knowledge, beyond their working lives, of visible minorities. What is clear is that no amount of community relations training is likely to redress the balance of a lifetime of socialization. As Gilroy (1987) reminds us,

> [i]t is fruitless, for example, to search for programmatic solutions to discriminatory police behaviour in amendments to the training procedure when professional wisdom inside the force emphasizes a racist, pathological view of black familial relations, breeding criminality and deviance out of cultural disorganization and generational conflict. If this racist theory is enshrined in the very structure of police work, it demands more desperate remedies than merely balancing the unacceptable content against increased "human relations" training. (P.109)

Discussion

The imagery of the alien violence and criminality personified in the "dangerous supplement" and the illegal immigrant has become an evocative technique in the hands of politicians and police officers whose authority is undermined by the political fluctuations of the crisis. For them, as for many working-class Canadians, there exists an irresolvable difference between themselves and the undesired immigrants as expressed in the latter's culture of criminality and inbred inability to cope with that highest achievement of civilization—the rule of law.

For most of us, television's overpowering images of Black deviance—its regularity and frequency—are impossible to ignore. Television news, with its focus on violent street crime, also fuels the stereotype of Black criminality. Many local news programs begin with a crime story often showing a Black man hunching over, shielding himself from the camera, being escorted away in handcuffs by police. Though Black women are not usually the focus of crime news, they too have recurring roles. They are frequently shown as battle-weary, grieving mothers, photographed crying over the death or arrest of their sons and daughters. These negative images have been seared into our collective consciousness. It is no surprise that most Canadians believe that Blacks are responsible for committing the majority of crime. No doubt, many of the suspects paraded across the nightly news may be guilty criminals. Here I do not intend to suggest that Blacks did not or could not commit crimes or to invoke a pastoral definition of the Black communities of the inner cities as places where crime does not occur. I seek instead to refer the reader to the images and representations of Black criminality, which seem to me to have achieved a mythic status in the lexicon of contemporary race politics. See also Gilroy's (1987b) arguments that concern the "mythical status of Black criminality." However, this is the myth what Russell (1998) calls the "criminal Blackman." On balance, the picture that comes to mind when most of us think about crime is the picture of a young Black man (Baker 1994). As a result there are several questions that can be raised: How is it that Blacks are widely seen as both symbols of success and deviance? What accounts for these conflicting perceptions of Blacks and blackness? At the core of the presentation of Blackness is the theme of contradiction. Images of the deviance in which Blacks are involved are promoted alongside images of Black achievement.

The picture of Black success is partly presented in the form of fictional media portrayals of Black and White friendships. In the film industry, for instance, Blacks and Whites are paired as partners-in-crime, blood brothers, co-workers, and friends. These images, however, do not effectively counter negative Black images. This is because sometimes underlying these Black and White pairings and images of Black success are subtle messages of Black deviance. The movies and television shows frequently portray Black achievement as a "reform" case. The television show *Designing Women* illustrates this quite well. The Black character, played by Anthony Bouvier, is an ex-convict hired as the gofer for four affluent Southern White women. In the movie *Shawshank Redemption,* two convicts, one Black and one White, become friends. In particular, the Black character, Red, was guilty of murder, while the White character, Andy, had been wrongly convicted of murder.

Another reason that Black success stories do not counterbalance images of Black deviance is that Black success is given a unique interpretation. In the minds of many, Blacks have "made it." In many instances,

Black superstars are not perceived in terms of their Blackness. A scene from Spike Lee's movie *Do the Right Thing* illustrates this point quite well. Two young men, Pino, who is Italian, and Mookie, who is Black, have a discussion about racism. Pino hates Black people and refers to them as "niggers." In an attempt to point out his racial double standard, Mookie reminds Pino that all of his favorite celebrities are Black (Magic Johnson, Prince, and Eddie Murphy). Pino responds, "they're not really niggers...They're not really Black, they're more than Black" (*Do the Right Thing* 1989). To some Whites, I would add, those Blacks who achieve large-scale success—Michael Jordan, Oprah Winfrey, Michael Jackson—becomes colorless. From this perspective, a credible measure of tangible progress without disturbing the basic class structure of society is presented. Those Blacks who conform to the criminal stereotype remain "Black."

Given this complex imagery, it is predictable that Blacks are viewed as emblematic of both success and deviance. The contradictory media representations of Blackness reflect a double-edged resentment—the threat of both Black crime and Black success. The result, as Russell (1998) argues, "is cross-wired thinking about Blacks and Blackness." While the media portrays neutral and, in some cases, positive images of Blacks, these images cannot compete with the overwhelmingly negative characterizations. This form of exclusion has larger implications for society, insofar as the widespread negative images of Blackness and crime raise several questions: What is the role of history in understanding today's criminal justice system? Does the criminal justice system discriminate against Blacks? Is there any way to improve the relationship between Blacks and the police? What role do the media play in fostering tension between racial groups? And, how can the law be used more effectively to address racial harms against Blacks? Any attempt to tackle these questions in relation to community policing *must* keep in mind the work of Antonio Gramsci (1971), Livy Visano (1994), and others that give useful insight and caution in the function of hegemony.

Hegemony

There are many interpretations of Gramsci's concept of hegemony. According to Lears (1982), Gramsci's reformulation of hegemony is the "spontaneous consent given by the great masses of the population to the general direction imposed on social life by the dominant fundamental group." Visano (1994) explains that hegemony is established through the "consent of the ruled, a consent secured by the diffusion and popularization of ruling class views of classes," the reproduction of "a prevailing consciousness [that] is internalised and becomes part of a 'common sense'" (p.200). Alternately, according to Boggs (1976), it is

> the permeation throughout civil society...of an entire system of values, attitudes, beliefs, morality, etc. that is in one way or another supportive of the established order and the class interests that dominate it...To the extent that this prevailing consciousness is internalised by the broad masses, it becomes part of 'common sense'. (P.39)

Similarly, hegemony could also be understood, as Simon (1982) posits, as "a relationship not of domination by means of force, but of consent by means of political and ideological leadership" (p.22). Sumner (1979) explains that the "ruling bloc has to subordinate the other classes to the requirements of the productive process not just by issuing decrees, but also through an ongoing transformation of moral values and customs in civil society" (p.26). Thus, civil society is the terrain on which classes contest for power and the battle ground upon which hegemony is exercised.

The definition operating herein combines all the elements that follow and which are related to the dispersal of power within structured belief systems and the consent which is given to the maintenance of a social totality. Within this context, the exercise of hegemony is infused into everyday practices through institutions such as the criminal justice system, the education system, organized religion, the family, folklore, the media, and the law. From a cultural perspective, these institutions are considered "neutral" to the extent that, according to Boggs (1976), they operate "without 'sanctions' or compulsory 'obligations', [yet they] nevertheless exert a collective pressure and obtain objective results in the form of an evolution of customs, ways of thinking and acting, morality" (p.40). However, the seeming neutrality masks the assertion of hegemony. For hegemony to be asserted successfully in any society, Boggs (1976) argued that it must do the following:

> Operate in a dualistic manner: as a "general conception of life"...and as a "scholastic programme" or set of principles which is advanced by a sector of the intellectuals...[Gramsci observed that where] hegemony appeared as a strong force, it fulfilled a role that guns and tanks could never perform...it encouraged a sense of fatalism and passivity towards political action; and it justified every type of system-serving sacrifice and deprivation. In short, hegemony worked in many ways to induce the oppressed to accept or "consent" to their own exploitation and daily misery. (Pp.39-40)

Consent

Consent is integral to the maintenance of hegemony and can be conceptualized as a series of power relationships geared toward the maintenance of a particular status quo. Consent accomplishes a social complicity on the part of the general population in the enforcement of norms. "Hegemony is the equilibrium," Visano (1994) suggests, "between leadership based on consent and domination based on coercion" (p.200). I might add that hegemony is a calculated mediation that operates within a liminal cultural space between frontiers and boundaries. However, as such, it is the working out of a "plot" that controls the contradictory space of disequilibrium that exists between "comfort" and the fear of its subversion by the "dangerous supplement." This artificial equilibrium and liminal space must be policed in order to maintain "the pleasure of meaning-giving order" (Rogin 1993:509). This consent can be achieved through force but it can also be achieved spontaneously. However, Gramsci's (1987) concern is with spontaneous consent, that which is "historically caused by the prestige (and subsequent confidence) which the dominant group enjoys because of its position and function in the world of production" (p.12). This consent does not occur in the same way as a political party gains the consent of the people by virtue of winning the most seats in an election.

Gramsci's consent is synonymous with consensus and, according to Lears (1982), is a "moving equilibrium." The concept of "moving equilibrium" is best understood in terms related to the emergence of counter-hegemonic views which are subject to co-optation and incorporation in the views of the ruling or dominant group as a means of re-establishing a new equilibrium or status quo. It is here that I will locate all commission of inquires. They facilitate and function to repair or re-established a sense of hope in the criminal justice system, like any other institution/system. By using the phrase "moving equilibrium," the implicit understanding is that hegemony, or the social production of consensus, is subject to a continuous shifting back and forth between the views held by the dominant group and those held by subordinate groups. The middle position or equilibrium is therefore a negotiated position, or a negotiated consensus. Gramsci's negotiated consent/consensus is viewed in relation to cultural practices, which are shared by both the dominant and the subordinate groups, for example, religious practices/rituals pertaining to birth, christening, marriage, and death. Likewise, commonalities in trading and arbitration practices become the basis for social interaction between the dominant and the subordinate groups.

Ideology

Ideology is also integral to Gramsci's definition of hegemony, but it is a concept not clearly defined. Here, again, Gramsci's meaning has to be inferred. Stuart Hall, Bob Lumley, and Gregory McLennan's (1978) interpretation links ideology to "superstructure" and they argue that it "cannot be understood outside the…structure/superstructure complex" of the Marxist conception of the state. They claim Gramsci's conception of ideology "is not judged according to a criterion of truth and falsehood, but according to its function and efficacy in binding together classes and class fractions in positions of dominance and subordination" (Hall et al. 1978:48). Further, these ideologies can "become compounded, serialized or clustered in ideological formations, such as conversational discourse, theory, law, theology and popular images" (p.49). Hence, there is not only one ideology, but also a complex of ideologies operating at any historical moment.

Louis Althusser (1971) offers the best explanation of the complex of ideologies in his essay titled "Ideology and Ideological State Apparatuses." Althusser (1971) develops a general theory of ideology, or what he termed a scientific theory to replace the descriptiveness of a Marxist notion of ideology as false consciousness. To be falsely conscious implies a "truer" consciousness, which is external to the lived reality of the subjugated individual; that is, a reality that is other than the dominant ideology. Althusser's (1971) conception of ideology as state apparatuses and ideological state apparatuses (henceforth ISA) are similar to Gramsci's notion of society as consisting of a base and superstructure. Althusser (1971) captures what Gramsci (1971) might have meant by cultural consent; that is, the accepting and "naturalizing" of certain "bourgeois" ideas and practices by "the masses," which occurs from the immersion of the individual within a matrix of overlapping ideologies as represented by institutions such as the church, school, family, law, media, the criminal justice system, and so on. Within these institutions, or Althusser's (1971) ISA, individuals acquire "know-how," "rules of good behaviour...rules of morality, civic and professional conscience which actually means rules of respect for the socio-technical division of labour and ultimately the rules of order established by class domination" (p.128). The dominant ideology within this matrix is the educational institution, which Althusser (1971) argued allowed for the "reproduction of submission to the ruling ideology for the workers and a reproduction of the ability to manipulate the ruling ideology correctly for the agents of exploitation and repression, so that, they too, will provide the domination of the ruling class "in words"...and in forms which ensure subjection to the ruling ideology" (p.128). Further, for Althusser (1971), ISA are both the stake and the site of class struggle and where "the role of the ruling ideology is heavily concentrated, the ideology of the ruling class which holds state power" (p.142). The ISA, particularly the educational ISA, are heavily policed by the "ruling ideology" and by ruling ideology, and we can assume that Althusser (1971) meant the dominance of one body of representation over another whether by overt or ideological means.

Within Althusser's (1971) social totality, individuals are "worked on /over" by the ideological structures amounting to what could be termed the ideological "marking" of the individual. This marking process, from a sociological point of view, is the equivalent of the process of socialization, with the "marking" occurring as an internal (unconscious) process that is manifested externally in practices. These ideological structures function to inculcate certain societal values and teach diversified and specific practical life skills geared toward the "reproduction of the means of production," which Althusser (1971) asserted are "provided more and more outside of production" (p.127). Further, this marking process cannot be avoided because, according to Althusser (1971), "individuals are born into and live through ideology" (p.27). It is a process of their conscious existence. Hence, ideology cannot be either true or false. Thus, an Althusserian conception of ideology accommodates the understanding of ideology as the acceptance of a particular status or organization of a civil society, and it operates at the level of "common-sense." For example, a belief in God, a belief in the need for education, a belief in the rule of law and so on; or, alternately, one that is perceived to be necessary and sufficient for the existence of any society that subscribes to "law and order" (as opposed to a state of anarchy or lawlessness). The latter conception holds to the notion that it is not necessary for there to be a prediscursive, truer knowledge existing somewhere that, once exposed, will make civil society less repressive. Further, even if there are those who are aware of how certain classes obtain their positions of dominance, such knowledge would not necessarily result in societal revolt.

Interestingly, there is a certain class reductionism in Althusser's theory of ideology, where all subjects are necessarily class subjects. However, as Mouffe (1988) argues, "Within every society, each social agent is inscribed in a multiplicity of social relations—not only social relations of production but also the social relations, among others, of sex, race, nationality, and vicinity" (pp.89-90). Mouffe's (1988) conception of society is a complex of heterogeneous social relations possessing their own dynamism and it is in this complex that the usage of ideology is linked firmly with symbols, the *meanings* they represent and the *functions* they serve. It is also the non-coercive part of hegemony, where the political state is bound together without force. Ideology (Hall et el. 1978) in this context therefore functions to "cement and unify the social block" (p.48).

Therefore, it can be argued that cultural consent involves more than just the activities of individual members. It involves collective interaction through daily rituals where it is ensured that people learn, practice, and reproduce the dominant ideology. This is similar to Bennett's (1992) reading of Foucault's conception of the "production and circulation of specific regimes of truth—regimes which organize the relations between knowledge and action in specific ways in different fields of social regulation" (p.32). It would therefore be significant for a critical criminologist and critical race theorist to take issues relating to ideology very seriously in that if we take Althusser (1971) seriously, then everyone, at some level, inadvertently reproduces their own oppression. Thus, it would be in our interest to embrace a concept of ideology because, as C. Wright Mills (1988) argues,

> the overturning of repressive social hierarchies requires not just the organization of oppositional forces, but the deployment of ideological infantries: the contesting of definitions, the challenging of presuppositions, [and] the recasting of concepts.

However, the creation of new ideologies must start from a clear understanding of how existing ones are manifested and how they function in society. Such an understanding is essential to the critical criminologist and critical race theorist in projects of creating social transformation, or, at the very least, for fundamentally undermining the legitimacy of the existing status quo.

The question that might need to be asked here is how is the law implicated in the practise of hegemony? As Vilhelm Aubert (1969) asserts, the relationship between law and power has long been noted: "[B]eneath the veneer of consensus on legal principles, a struggle of interest is going on, and the law is seen as a weapon in the hands of those who possess the power to use it for their own ends" (quoted in Smart 1992). The following section therefore highlights the linkages between political and legal hegemony, which tend to be omitted from race analyses.

Hegemony and the Law

In most societies, the law, even narrowly defined as an institutional system of jurisprudence, should be seen as constituting a principal vehicle for securing the hegemony of the state. Law shapes, and is shaped by, the state's socioeconomic interest. The general activity of law is, however, much wider than state or governmental activity to the extent that, according to Gramsci (1987), it includes the "activity involved in directing civil society in those zones...call[ed] legally neutral, [that is] in morality and in custom generally" (p.159).

The law exists in the intersection between the dominated and the subordinated groups; in the places where the consent of the state is negotiated. This mandate has remained basically intact since the early nineteenth century. As Gramsci (1987) tells us that, "protecting and maintaining the boundary between public and private, or alternately, between private property and public power" (p.23). Not only does law maintain this boundary. Mensch (1982) asserts that the law is "able to define and separate the private sphere and private autonomy while simultaneously defining zones where public power can be exercised freely and absolutely" (Mensch 1982:23). At the point where the public and the private spheres meet, the function of the law can be ambiguous in that it can either serve the purposes of the dominant group or the purposes of the subordinate group, alternately or simultaneously. This ambiguity also contributes to its elevation to a privileged and hegemonic position. As noted by Sumner (1979),

> [t]he legal system is founded upon a series of ritually articulated ideologies which work to the benefit of the dominant class...[and] is but one of several ideological forms which combine to form and reproduce the ideological kernel of class hegemony. (Pp.4-9)

Legal hegemony is intertwined with political hegemony in Western liberal societies and, in a "modernistic" way, diffuses its worldview onto subordinate groups such that those groups will adopt a legal view as "natural." The hegemony of law is thus dependent on people believing in the necessity for legal structure, particularly if they adhere to liberal notions of civil society. Legal hegemony is also dependent on its access to the punitive (repressive) institutions of the state's apparatus of social control such as sending someone to prison if he/she breaks the law or imposing fines, monetary and otherwise. This equates with Gramsci's (1971) conception of the law as being "the repressive and negative aspect of the entire process, civilizing activity undertaken by the State" (p.287). It also approximates Michel Foucault's "regime of truth," wherein, according to Foucault (1979) the legal apparatus, in both everyday practice and the criticism of institutions, one sees the emergence of a new strategy for the exercise of power to punish...[and] to make of the punishment and repression of illegalities a regular function, coextensive with society; not to punish less, but to punish better; to punish with an attenuated severity perhaps, but in order to punish with more universality and necessity; to insert the power to punish more deeply into the social body. (Pp.81-82)

The law is legitimized through consent, coercion, and public opinion. Consent herein is derived largely from the idea that legality provides the basis for community living. To the extent that people obey the laws of society is tantamount to ratification or consent even if there may be specific laws that are problematic. Coercion here can be the direct state control required to enforce particular laws or it can be coercion linked to the private sphere and the withholding of particular goods or services with the intent to punish. Public opinion is also that which is linked to attitudes, which are designed to sanction non-conformist behaviours; the utilization of which results in public acceptance of what constitute the truth. This is the law at its most hegemonic stage—legitimatizing the state by aiding in what Barrett (1991) tells us is the "organization of consent—the processes through which subordinated forms of consciousness is constructed without recourse to violence or coercion" (p.54). It is precisely at this point that Black communities are sold social peace at the cost of social injustice.

It should be reiterated that I have been arguing that the functioning of political and legal hegemony are essentially the same, and that the systems of domination, which are erected by both inadvertently or deliberately and are designed to keep women, for example, in the private sphere (at home), are basically the same. Legal and political hegemony are also synonymous with Eurocentrism, which, from a Black perspective, developed stereotypes of Blacks as dependent, domestic, emotional, and sentimental. This "ideology" that Blacks are the "inferior race" is what Nunn (1997) calls "the Eurocentric cultural experience" or "laws which maintains White cultural hegemony through false universal claims and the privileging of the White historical experience" (p.358). To this end, the law should not be perceived as being benign or neutral; not by anyone and, least of all, not by Blacks.

Culture

One of the means through which hegemonic order achieves its desired results is culture, which pervades social relations and is both a product and a producer of ideological values and institutional systems. As a construct of power relations, culture functions to solidify and perpetuate power through the inculcation of media images and "common sense" beliefs. Visano (1994) comments: "Culture, organized on hegemonic principles, is crucial to the political economy, which in turn depends on instruments of authority to maintain control over capital and profit" (pp.195-196). Moreover, culture drives the imperial frontier forward while insulating it against subversion.

However, there persists an interdependency between policing and criminality similar to that between constructing an ontology of "sameness" ("us") out of "otherness" ("them") The frontier of order only exists on the basis of the non-order that the power lines of hegemony demarcate, the linear divisions from which deviancy may be constructed, negotiated, and submitted to surveillance. However, the violent "cut" of the frontier is also the margin at which the disordering role of the counter subversive is enacted, hence criminalization.

Criminalization

If order is based on the disordering function of relations of warfare, then, as Foucault (1977) suggests, "penal justice...is intended to respond to the daily demand of an apparatus of supervision half submerged in the darkness in which police and delinquency are brought together" (p.282). He proposes that during the nineteenth century the "direct, institutional coupling of police and delinquency took place: the disturbing moment when criminality became one of the mechanisms of power" (p.283). Thus, the police make use of classifying delinquency to control illegalities and to create the criminal as a commodity of a control cercarial industry.

A further connection was established, however, between policing and racism and criminality. At the same time as the panoptic system of penal justice created "docile bodies" of the public, the colonial legacy of racism was utilized to differentiate and simultaneously assimilate "non-Whiteness" into a criminogenic, racialized category of counter-hegemonic threat. Thus, while the system could isolate and differentiate illegalities and individualized delinquency, the policing function could also demonize and imitate the hegemonic construct of deviancy by representing itself as an army defending the boundaries of delinquency, which are brought together specifically in the constructed "darkness" of ghettos and inner-city "jungle-ized" colonies. Within this liminal landscape, the police deploy counter-subversive invasion tactics to combat the "monster" of the "otherness" they first produce and then cannibalize as a means of generating power.

The creation of what Hall et al. (1978) term a "sub-proletariat" or "super-exploited" class of Black workers within British cities is actually an internalized re-colonization of immigrants from former colonies. Thus, marginalized urban areas of poverty and high unemployment tend to be transplanted "colonies" of previously conquered and subjugated peoples for the purposes of providing a reserve army of labor (just as people of color were reserve armies of labor in the former colonies). Although accepted as immigrants into the original colonizing nation (Britain), the sub-proletariat underclass, according to Hall et al. (1978), continues to be the colonized "other," and is treated educationally, socially, and legally in a paternalistic manner as if its members still remained the "occupied" populace of an imperial possession (pp.362-381). This situation is also apparent in Canada, as the criminal actions against West Indian suspects in crimes suggest (for example, the "Just Desserts" shooting or the highly publicized killing of a police constable in a drug-related incident in the Jane-Finch corridor). The individuals involved are painted as criminal aliens who, despite having lived in Canada since they were children, brought their "disease" of criminal "otherness" with them from the "disorder" of the distant colonial territory.

In the United States, the original importation of Africans into the thirteen colonies for the purposes of slave labor has led, I suggest, to a similar situation where Black citizens are treated as an alien, disorderly presence. Even though "freed" from slavery, Afro-Americans are still represented as bearers of "jungle" behavior and mentalities, which include propensities toward crime and drug use, into white America.

Since the start of the twentieth century, anti-drug legislation (in the United States beginning with the passing of the Harrison Act in 1914) has been utilized to engage in surveillance on suspected "dangerous" elements of the population and to segregate citizens on the basis of race and class. Historically, fears of counter-hegemonic forces were situated geopolitically, as drug use was alleged to be concentrated among the poor, immigrants, and racial minorities. Thus, the drug threat was constructed as being the product of foreign, alien influences as well as being situated within the peripheral regions of urban decay and social dissolution.

Black "colonies" are ironically and liminally located in the urban core of cities while White middle-class families have moved to the "peripheries" of suburbia—escaping, perhaps, to the "pastoral-like" promise of newly made frontiers—once more taming "nature" as a continuing process of modern "civilization." Infantilized as a sub-social "race" incapable of self-management, the Black "colony" of contemporary EuroAmerica has been additionally marginalized by abstract notions of hegemonic family values. Ignoring economic and social conditions whereby most Black males are denied employment, often the government's response has concentrated on the proliferation of Black families led by single mothers and has pinpointed the inherent "weakness" of Black family structures as a further etiology of disorder, drug abuse, and criminal production (Solomos and Rackett 1994; Staples 1990). As Armstrong (1994) argues, "the result...is a single melodrama in which 'the family' becomes the victim of crack. This practice invites us to hold specific but nameless individuals responsible for destroying a highly abstract set of roles. The war against drugs is transformed into a war against the women and children who are victims of drugs" (p.23). And, as David (1991) asserts, anti-drug strategies have served to target Black women as producers of criminogenic children who are biologically "programmed" to become gang members, drug dealers, and drug addicts (pp.182-183).

Never accepted as members of a regular proletariat, these Black inner "colonies" of the descendants of former African slaves and of more recent West Indian immigrants are kept predominantly underemployed and forced into adopting the role of a new lumpen proletariat. As the latter, they are often compelled to choose street hustling as an alternative means of petty capitalistic survival (Hall et al. 1978:362-81).

The nuclear family unit and standardized suburban household permit discipline and control of the general population, of the "super-exploited" into a panoptic classification of illegality. Furthermore, drugs have become a device of double-colonization: Firstly, as a means of colonizing the poor and marginalized of the Black community—through addiction and dependence on imitative and commodified capitalist employment/enslavement by the illegal drug industry—and secondly, as commodities of a cercarial system of police surveillance whereby criminal drug use can be utilized as a means of listing, differentiating, isolating, and policing a dangerous and racialized underclass.

It is important to recognize that a significant section of Black communities live on the margins of "official" society and hardly fit, therefore, the vision of an urban proletariat. Living in and around Black communities, their presence, at the same time, is a powerful reminder of (a) the precariousness of the material security of the Black communities, and (b) the alien standards of judgment that are often deployed against them to "manners," domesticated and marginalized them. However, while this serves to permanently exclude and deny these marginal sections, it also provides a freedom and mobility that opens them to a range of discursive possibilities.

A Concluding Remark

Permanent unemployment/underemployment, marginalization, and the criminal depictions often provided by the mass media and sustained by the general public's belief that most Blacks are criminals produces a certain criminality. This phenomenon tends to mark entire communities, resulting in exclusion. And it is this specter of criminality that haunted middle class Toronto. The emergence of Jane/Finch and Regent Park as the sites of this criminality served to not only telegraph avoidance but also reveals the significant moment these communities were established as the "dangerous supplement." And once these communities were so named and fixed in the imagination of all, the mechanisms of exclusion and social control are inescapable. In other words, the "criminality" of the communities creates a "moral panic" for middle-class Toronto and by extension the prevailing hegemony which produced its own "ideological frame" and subsequent responses to contain, domesticate, and eventually control. Herein lies the call for community policing. It is a call for a form of policing that appears to give Blacks the power to define what "order" should be policed. It is a call for community policing to maintain "order" out of "disordering" Black communities, and, thus, a form of exclusion.

Chapter 4

Race, Citizenship, and Exclusion

This chapter will examine the utility of the concept of citizenship in understanding both the ideological and material conditions under which Black people live their lives in Canada, and under which they receive discriminatory treatment within the criminal justice system. Citizenship is essentially about inclusion and membership. It will be argued that for Black people in the twenty-first century formal and informal mechanisms of exclusion serve to deny them their full social, economic, and political rights of citizenship. Exclusion manifests itself through the denial of legal citizenship (effected through immigration policies) and the denial of social citizenship through the experiences of poverty and racism. A case study will be provided. After a brief discussion of the terminology of citizenship, the concept will be analyzed with particular reference to the legal and social policies that both generate and reinforce exclusion.

Marshall (1950) outlined three elements that constituted citizenship: the civil, the political, and the social. The civil element is associated with the rule of law: liberty of the person, freedom of speech, and the right to justice. The political element concerns the ability to participate in the electoral process as a representative or as a voter. Thirdly, the social element consists of a whole range of activities, from the right to economic welfare and security to the right to share fully in the social heritage and to live the life of a civilized being according to the standards prevailing in the society (Marshall 1981:10). The realization of full citizenship, as Cook reminds us, "involves three different sets of social institutions: the criminal justice system, parliamentary institutions and those institutions concerned with the provision of education and social welfare" (Cook 1993:137). However, their contributions to citizenship are, to a degree, interdependent. As Barbalet explains, "a political system of equal citizenship is in reality less equal if it is part of a society divided by unequal conditions" (Barbalet 1988:1).

Implicit within citizenship theory is the assumption that citizenship itself has the capacity to erode such inequalities. In his earlier works,

Marshall believed that "citizenship may unite a society divided across class lines and impose modifications on class" (1950:84). However, thirty years later, he argued that "the right of citizenship inhibits the egalitarian tendencies of the free economic market, but the market and some degree of economic inequality remain functionally necessary to the production of wealth" (1981:12). As Cook argues, "citizenship may provide a basis of *apparent* equality upon which the structure of capitalist inequality could be built" (Cook 1993:137). It is clear then that citizenship is seen as capable of uniting where class divides, but the apparent unity gained through equal legal and political citizenship rights serves to conceal the fundamental economic inequalities inherent within capitalist societies.

Leaving the functionalist arguments aside, the critical issue here is that full and equal citizenship, involving the elimination of social inequality, appears incompatible with the operation of free market economics. It is useful to point out that Marshall's (1981:11) notion that "all who possess the status are equal with respect to the *rights* and *duties* with which the status is endowed" fits neatly within the conservatives' reworking of the concept to emphasize the "duties" of citizenship. This theme is central to the responsibilization strategy, one of the many salient mobilization technique for community policing. Within this perspective, the "entitlements" of citizenship are conferred only upon those who are responsible and active in carrying out their duties as citizens.

For theorists on the left, the right of citizenship remains the paramount concern. As Lister (1992) explains,

> [c]itizenship is defined in social as well as political and legal terms: it denotes the ability to participate fully in the social and political life of the community…Poverty is corrosive of citizenship…For women and members of Black and minority ethnic communities living in poverty, the exclusion from full citizenship is often compounded. (P.vii)

It is useful at this point to provide a brief history of Caribbean immigration in order to contextualize the discussion. Caribbean immigrants began entering Canada at the turn of the twentieth century. They came primarily to work in the steel mills and coal mines of Cape Breton, Nova

Scotia, and, around 1911, they arrived in the Province of Quebec as domestic laborers. The shortage of domestic laborers prompted the emergence of the first Caribbean domestic scheme with Canada (Calliste 1991). Thus, between 1922 and 1931, three quarters of the Caribbean Blacks who came to Canada were domestic servants. However, many were soon deported because of an imminent fear that they would become a public burden. Nevertheless, Macklin (1992:688) notes that the deportation coincided with periods of recession; thus, Blacks were fired to make room for White servants.

In 1955, a second Caribbean domestic scheme emerged. This scheme resulted from the lobbying efforts of the Caribbean government and Canadian employers. The former accused the Canadian government of having racist immigration policies, which deterred the entry of Caribbean immigrants into Canada. Canadian employers, on the other hand, fought for the scheme because Caribbean domestics proved to be cheaper labor than other groups. Consequently, in 1955 an agreement was reached between Canadian employers and the governments of Jamaica, Barbados, and Canada (Macklin 1992:689; Calliste 1991:105): "Under the Scheme, single women between the ages of 18-40 with no dependents and at least an eighth grade education were admitted to Canada as landed immigrants on condition that they remain in live-in domestic servants for at least one year" (Macklin 1992:689).

Both Macklin (1992) and Calliste (1991) commented that the decision to admit Caribbean domestics as landed immigrants came about primarily because of two factors. They noted that the first reason was in response to the criticisms that Canada received concerning its exclusionary and racist immigration policy, which many likened to an indentured labor protocol. The second reason was the belief held by both Canadian employers and Canadian governments, that unlike European immigrants, Caribbean immigrants were less likely to leave domestic service at the end of the one-year period. As well, Macklin (1992) and Calliste (1991) noted that the decision by the Canadian Government to establish the 1955 scheme also came about because the former retained the right to deport a woman, at the expense of Caribbean governments, in the first year if she became pregnant or terminated her contract with her employer (Calliste 1991:106; Macklin 1992:689). Despite this, however, Macklin (1992:693) observed that under this domestic workers program, 70 percent of the Caribbean immigrants who applied for landed status were successful.

Increasingly, the Domestic Workers Scheme came under attack. It was criticized as being unnecessary in lieu of the proposed changes to Canada's immigration policy. These changes, which placed an emphasis on skills as the main selection criterion for immigrants, occurred in 1962. In addition, the Caribbean Association in Ottawa noted that the scheme condemned Caribbean women to a second-class status in Canada and should therefore be cancelled. Alternatively, the Association suggested that they should be treated as skilled immigrants under Canada's "preferred" immigration policy. As Hawkins (1991:27) points out The Preferred Immigration Policy was established in 1993. It allowed Canada to select immigrants from preferred countries to fill certain occupational positions in Canada. These countries included Britain, Norway, Sweden, Denmark, Finland, Germany, Switzerland, Holland, Belgium, and France Robinson (1993).

Furthermore, Caribbean domestics were reportedly being paid less than their White counterparts. Some critics of the scheme estimated this wage difference to be about $150 per month. In addition, others complained that the small number of Caribbean immigrants admitted under the scheme did little to curb the increasing unemployment rate in the Caribbean (Calliste 1991:106-113).

Caribbean governments were also criticized for their involvement in the scheme. It was argued that Caribbean governments fostered the "brain drain" from the Caribbean. Under the scheme, in order to maintain favorable relations with the Canadian government, Caribbean governments selected people on the basis of skills and education rather than past employment in the domestic service. This practice was further aided by the changes to Canadian Immigration Policy in 1961, which required that the educational requirements for domestic workers from the Caribbean be changed from grade eight to grade nine (Calliste 1991:106-113). As a result, "many women were qualified as teachers and secretaries, but came as domestics because other avenues of immigration were closed" (Bolaria and Li 1988:201). World events, as explained below, along with the introduction of new immigration policies from the 1960s onwards led to an increase in the numbers of Caribbean immigrants entering Canada.

First, the new policies were aimed at removing racial discrimination, but instead established skills as the main criterion in the selection of unsponsored immigrants. Consequently, Caribbean immigrants were not able to apply for permanent status in Canada other than through the domestic workers scheme. Secondly, in 1962 Britain ended its open-door policy between itself and its Commonwealth dependencies. As a result, Canada and the United States became the principal countries of choice for Caribbean emigrants (Anderson 1993:38). Thirdly, countries such as Jamaica and Trinidad and Tobago got their independence in 1962. Hence, these newly independent Caribbean islands could now conduct diplomatic relations with other countries, as well as negotiate mutually agreed upon immigration policies. Fourth, Caribbean migration during this era was further encouraged by Canada's need to rebuild itself after World War II. As a result, in both Canadian and American societies, there was a need for middle and upper-level skilled workers: "The Caribbean was over-supplied with skilled workers, who because of the underdeveloped state of their own economies, were forced to leave their own homeland" (Anderson 1993:37).

Between 1962 and 1967 the number of landed immigrants from the Caribbean arriving in Canada was still under 5,000 per year. Many critics maintained that the new regulations were still racist, as they retained policies that favored the sponsorship of a wide variety of relatives for European immigrants. This privilege was not extended to non-European immigrants (Hawkins 1991:38-39). Interestingly, the new Immigration Regulations of 1962 sought to eliminate all discrimination based on race, color, or creed. In practice, however, the Immigration Regulation still favored Europeans. Thus, Canadian officials "rightly saw that Canada could not operate effectively within the United Nations, or in the multiracial Commonwealth, with a millstone of racially discriminatory immigration policy around her neck" (Hawkins 1991:39).

In 1970, the United Nations designated 1974 as "World Population Year." This made population growth policy a high priority on Canada's national agenda. The aim of this UN declaration was to encourage countries to link population growth with economic and social development. One way Canada had traditionally sought to do this was by allowing more immigrants to enter Canada, and 1974 was no exception. Thus, one of the consequences of this announcement was an increase in the number of

Caribbean immigrants entering Canada in that year (Hawkins 1991:44). In a legal sense, citizenship was, and essentially still is, a device for the administration of immigration control. After a careful reassessment, key themes do emerge from the historical and political literature, which help to explain the ways in which the status of citizenship is still formally and informally denied Black people. One such theme is the racialization of immigration and crime issues in Canadian society.

Bill C-44 illustrates this quite well. In this respect, Bill C-44 was an enormous shift in the Immigration Act, particularly in light of the immediate and future consequences of deportation. Section 70(5) deprives individuals in Canada of their rights to various avenues of recourse from the issuance of a dangerous opinion and their corresponding removal from Canada. This Bill affects permanent residents regardless of the length of time they reside in Canada. Once convicted of certain offenses they would be deported. Section 70(5) removes the right of a permanent resident to appear before the Immigration and Refugee Board (Appeal Division) and appeal such a deportation order with respect to the alleged illegality of their removal order or the possible inequity of removing them from Canada. When an Applicant has been determined to be a "danger to the public," sponsors (Canadian citizens sponsoring a member of the family class for permanent residence in Canada) also lose their right of appeal pursuant to subsection 77(3.1).[3]

In light of the insidious nature of the immigration legislation of Bill C-44, it is useful to delve into the social and political context that beset its introduction and immediate implementation against persons in Canada. In January 1994, Sergio Marchi, the Immigration Minister at the time, announced a public consultation process aimed at setting immigration targets for the next ten years (Noorani and Wright 1994:29). Anti-racism activists received this announcement with skepticism; they felt that the Reform Party's influence and a prevailing anti-immigrant climate were behind the government's action (p.29). The former Tory government, it must be remembered, had introduced three major bills on immigration and refugee policy, designed to restrict the number of new arrivals and restrict the appeal rights of refugee claimants (p.29), a theme common in Canadian immigration policy. Further, there was a global shift to the right and Canada, like many other countries, began closing its doors to immigrants and refugees. Noorani and Wright (1994) comment on this global shift toward conservativism and the trends against immigration:

> In Europe far-right political parties were helping to frame debates on immigration issues. In Germany, Neo-Nazis torched refugee shelters and immigrants' houses and the government responded by tightening its refugee policy. In Canada, white supremacists groups began to grow in urban centres across the country.

This political reality and social climate had a profound impact on the shift in the Liberal government's immigration policy. As Noorani and Wright (1994) point out,

> [t]he rise of the Reform Party and the recession inspired increase in hostility towards immigrants and quickly drove the Liberals to abandon their Trudeau era image as the immigrants' party. After coming to power, the government's rhetoric moved steadily to the right, in accordance with the prevailing anti-immigrant political attitudes and they refused to take the political risks that being pro-immigrant entailed.

Another factor that influenced the government's immigration policy was the "perception" by Canadians "that a significant number of crimes were being committed by immigrants," a perception shaped, in part, by the mass media. However, two particular cases elucidate this further: The homicide death of twenty-three-year-old Georgina Leimonis on April 5, 1994 (*Toronto Life*, 1994, p.7) and the shooting of Todd Baylis, a police constable killed in the line of duty (Cannon 1995:9).

The death of Georgina Leimonis came about during a late night robbery of the *Just Desserts* Restaurant in Toronto. The outrage that followed Ms Leimonis's death created a climate of considerable delirium. Torontonians said they no longer felt safe in their city, though statisticians and criminologists alike were saying that the crime rate for reported crimes of a non-violent nature had actually fallen and the reported rate of violent crimes had remained stable for years.[4] Similarly, Todd Baylis was shot and killed in Toronto in June 1994, by Clinton Gayle, who had been ordered deported to Jamaica in 1991 (Cannon 1995:10) This shooting by a deportee less than two months after the shooting of Georgina Leimonis became a point of reference for anti-immigrant lobbyist groups in linking crime to immigration.[5] The political and the social climate fostered the perception of a connection between immigrants and crime. In such a political climate, the Liberal government responded to this perception by embarking on a political campaign to rid Canada of its "undesirables" and to improve its image of being tough on crime.[6]

The public frenzy prompted the formulation of a special Royal Canadian Mounted Police task force to track down and deport foreign-born criminals. This also triggered the implementation and passage of Bill C-44. It must be remembered that the campaign to "clean house" was launched in July 1994 following the death of these two individuals. Many lawyers, legal scholars, and organizations have condemned the outrageous nature of this politicization of immigrants and crime. Critics noted that the government's conflation of immigrants and criminality as synonymous were used for political gains.[7] However, the position of the government was a platform of the Reform Party, a position that had led Preston Manning to a hairsbreadth away from becoming the country's Official Opposition (Cannon 1995:1).

The perception of criminality attached to immigrants led to renewed calls by Bishop Sotirios Athanasoulas, Head of the Greek Orthodox Church in Canada, for the collection of crime statistics based on race and country of origin, demands for greater police powers, more rights for victims, and less concern for the rights of criminals (Cannon 1995:11). Interestingly, here the figure of the "other" is combined with an intriguing development of spatial metaphors in Canada's law and order discourse. Crime is constructed as a threat to "inner security," a term which embraces the emphasis by the Justice and Immigration Departments on the protection of the legal order and, simultaneously, individual feelings of confidence in the ability of law to ensure order. Questions about "inner security" are, thus, regularly used to establish individuals' psychological sense of security and well-being. Their perceptions of safety levels within the communities, of personal safety, and, more narrowly, of fear of crime became manipulated. It is clear, then, that the appeal of the Bishop to "inner security" plays a role in Canadian criminal justice discourse. Thus, the fear of crime strikes at the heart of a central promise held out by the legal, political, and constitutional order, arguably leaving Canadian society vulnerable to unmanageable moral and, perhaps, psychic panics in situations of a rapidly increasing fear of crime. This, in turn, raises an interesting set of questions about whom such panics have generally been directed against? As argued previously, it is part of the role of the mass media to shape perceptions into seeing and non-seeing.

The media may have had a significant role in shaping public and political opinion, which led to the passage of Bill C-44. Following the deaths of Ms. Leimonis and Constable Baylis, the Canadian public was overwhelmed with stories linking crimes with immigrants and the flaws in immigration policies.[8] The *Toronto Sun* ran what many felt was a racist, irresponsible, and sensational campaign playing into the perceived link between crime and race.[9] The media did not portray the shootings as isolated incidents. Rather, the media portrayed entire Black communities,[10] particularly immigrant Black people, as communities of criminals. The publicity surrounding these two shootings helped obscure that the suspects in the Leimonis case were Jamaicans only by nationality, having spent most of their formative years in Canada. Interestingly, the suspects—who were characterized as "tough, armed, dressed for court in their flashy clothes, grinning at the cameras, proven lechers, with their housefuls of fatherless children" (Cannon 1995:14)—generated a media frenzy and public outcry, yet there was no such media or public outcry directed at the heinous crimes committed by other immigrants. For example, a White German immigrant, Wolfgang Muehlfellner had killed his wife, burned her body, and buried the remains in their backyard. He was later convicted of manslaughter in 1985 and ordered deported in 1988. He appealed the ruling and was allowed to stay in Canada until 1994, when his deportation order was stayed (African Canadian Legal Clinic 1997:9).

The social climate, political landscape, and media hysteria over two high-profile deaths had set the stage for the passage of Bill C-44 and the institution of Section 70(5). As the *Canadian Press Newswire* acknowledged,

> [l]egislation that aims to keep criminals out of Canada—or make it easier to kick them out if they get in—received royal assent…the last legislative step needed to become law. Bill C-44, sponsored by

immigration Minister Sergio Marchi, will bar refugee claims by people convicted of serious crimes, tighten parole rules for those under deportation orders and toughen the appeals process against deportation. The legislation was said to be a response to public outrage over cases like that of Oneil Grant, charged in the highly publicized 1994 slaying of a Toronto cafe customer [Georgina Leimonis].[11]

This legislation led to a further erosion of the demarcation lines between immigration and criminal law. Tannis Cohen (1988) refers to law as "a two-edged sword that can be wielded to further justice or to persecute and oppress." In the context of racism and law, this statement implies that although law has the ability to protect against racism and foster equality, law remains a principal tool for maintaining racial inequality. However, as illustrated from Canadian Immigration laws and policies, I postulate the following:

- People of European ancestry have been and continue to be the "preferred race"—always welcome in Canada
- Racism is inherent in the Canadian legal system, manifesting itself in both neutral and overt forms
- People of non-European ancestry—the "non-preferred" races—are excluded and rejected from Canada

David Matas (1989) sums up the relationship between law and racism in Canada's immigration system in this profound statement: "to talk of racism *in* Canadian immigration policy…is being generous. Rather we should talk of racism *as* Canadian immigration policy" [emphasis in original] (p.29).

The intersection of race and law has been, and remains, a fertile ground for much debate between critical, race, and legal scholars.[12] Basically, these scholars represent two schools of thought: the first school is composed of Marxists, Neo-Marxists, and some civil rights advocates. This school asserts that the law was established and exists to advance the interest(s) of the dominant group or class, deliberately ignoring the oppressive effects on people of color. The second school, championed by liberal theorists, posits that law is an appropriate medium for contending, advocating, and obtaining rights against racism. However, both schools reveal the contradictions inherent in the law. This book, tilts predominantly in favor of the first school of thought, but also conceives of the second school as a vehicle for effecting change within an inherently biased legal order.

In Canada, as elsewhere, the history of so-called modern legal institutions depicts law as an instrument used to establish, maintain, and justify the dominance of racial "elites." The logic of law underlies the institutionalization of Eurocentrism[13] and was used by a nascent capitalist class to establish its hegemony. By obscuring the inequalities created by European domination, law protected and ensured the privileges of the dominant group while serving as an instrument to control and exploit the majority. Law was used not only to exploit and to perpetrate domination, but it also served to justify European domination.

In the contemporary arena, civil rights activists have made significant inroads in criticizing and seeking redress for these skewed applications of law. Despite these efforts, the law is still stacked, not only because it constructs a world too colorless, but more so because it expressly disfavors people of color. Steele (1993) examines the prevalent myth of a color-blind nation. Steele asserted that persons with political, social, and economic power, "including those who manage to obtain a free ride because if their skin color" (p.599), continuously promote the notion of color neutrality within a nation that is blended by various races. This fairly accurate perception of the color-blind nation extends to the Canadian legal system.

It is clear that the government used the political landscape, social climate, and media hysteria surrounding two isolated crimes in Ontario to justify regressive changes to immigration law and policy. The government's policy under Section 70(5) sought to classify individuals as mere immigrants, rather than to focus the substance of Canadian residents' long-term relationships with Canada. The result has been to force out of Canada individuals who are *de facto* Canadian citizens. The policy does not affirm the notion of fundamental equality of citizens and non-citizens before the law. Rather, this legislative policy reinforces inequality and unequal treatment of Canadians residents in both its implementation and impact.

To summarize, through a discussion of the principal themes underpinning Canadian immigration policy, the argument in this chapter is that the legal and subjective citizenship of Black people in Canada has always been regarded as questionable and problematic. Although, the very word "immigrant" may now appear dated and inappropriate, the racialized use of the term persists and has an impact upon the everyday experience of Black people in Canada. Regardless of their legal rights and places of birth, black Canadian citizens often find themselves regarded as "alien," formally within, but informally without, citizenship. In addition, other vocabularies are developing which serve this same function of exclusion. The terms "refugee" and "asylum seeker" are no longer positive epithets but are now increasingly being connected within an essentially punitive discourse around "dubious claims" to welfare and affordable living.

The Civil Element

Discussions on the theory of citizenship often focus on a series of positive rights, which grant freedoms and entitlements such as the rights of living, justice, and political rights (including standing for, and voting in, parliamentary elections). However, when connected with the issue of race, it is also important to broaden any discussion of the civil element of citizenship to stress "negative" rights. These include the right of redress and complaint, which is the essential element in the Public Complaints Commission, and, of more importance, to include a Black person's right to walk the streets unencumbered by police or racist attacks. The state is quite slow to respond to the threat that racist attacks pose to this most basic freedom. It would be a mistake to write off all racist attacks as the work of fascist organizations, though they may play a part in creating a climate that encourages racial violence.

Against this background, it seems that for many Black people in contemporary Canada even the negative rights of citizenship—that is, the right to redress, the right to live in the communities without fear—appear illusory. Thus, the civil element of citizenship cannot, to this extent, be separated from the political and social.

The Political Element

According to the 1996 Census by Statistics Canada, those who identify themselves as ethnic minorities account for 16 percent of Toronto's population of 4,232,905 at the time of this research. Interestingly, just as women (who account for 51 percent of the population in Ontario) are grossly underrepresented with approximately 22 seats in a parliament of 130 seats, so are Black people. However, counting the number of Black MPs is by no means an adequate measurement of political participation. In regards to the primary right of political citizenship, that is, the right to vote in elections was removed in 1985. In 1985, a Municipal Act was passed that effectively eliminate all landed immigrants from voting in Municipal Elections. This Act became effective in 1988. Previously all immigrants or British subjects were allowed to vote at this level. See R.S.O. 1996: Chap. 32. Sec. 17 (b) (1). As Lea and Young (1984) have argued that Black people "are marginalized both economically and politically...that political marginality is not merely about possession of the vote but, is above all the exclusion from the ability to form coordinated, stable interest groups able to function in a process of pressure group politics." Gilroy (1987) also offers a perspective that helps us understand this political element. He suggests,

> [i]t is not that Blacks lack the means to organize themselves politically but that they do so in ways which are so incongruent with Britishness that they are incapable of sustaining life! Their distance from the required standards of political viability is established by their criminal character. Thus black crime and politics are interlinked. They become aspects of the same fundamental problem—a dissident black population. (Gilroy 1987:117)

This discussion also applies to the Canadian context, because it concerns not only alternative conceptions of politics, but also the notion of *Canadian-ness*, a theme echoed throughout this book.

Avoiding the danger of complete relativism, it is possible to argue in light of this discussion of racism, immigration and citizenship, that the struggles of Black people are more likely to be constituted as more criminal than political. This is possible not only because of the powerful "myth of Black criminality" (Gilroy 1987), but also because those involved in such struggles are ideologically constructed as always excluded from full and active citizenship, which by definition includes political participation. For Black people, the political element of citizenship is contingent upon the legal element. Thus, the experience of political participation is similarly limited by and constructed within the racialized discourse of exclusion.

The Social Element

Poverty and the stigma associated with the claiming of welfare benefits have long been associated with exclusion. For welfare benefits claimants in general, social exclusion operates at different levels. Ideologically, through a conservative discourse, the poor are stigmatized as responsible for their own failure. This is most clearly articulated in the Ontario's Government *workfare* plan. It is a divide that exists between the enterprise culture and the benefit culture. An ideological divide has been created which excludes the poor. This constitutes the use of guilt and shame, essential components of the claiming process, both in theory and in practice. In reference to the neo-conservative re-working of citizenship theory, Lister (1990) persuasively argued that

> [t]he New Right focuses upon the duties and responsibilities which citizenship demands rather than the positive rights with which it is endowed. They thereby seek citizens who are independent (that is, working!), active (giving to charity, doing voluntary work in their spare time), but lurking behind the active citizen is the successful, self-reliant, enterprising citizen, alias the consuming, and property-owning citizen. The unsuccessful and un-enterprising are thereby excluding from the ranks of citizens. (P.15)

If social citizenship is defined in terms of equal access to education and social welfare, it can be argued that welfare rationing and institutional racism serve to deny many Black people their full citizenship rights. Immigration policy has profoundly influenced social policy. Physical exclusion, by means of tough immigration controls, is accompanied by social exclusions, often through the welfare state. Even when Black people formally possess citizenship, they are dispossessed through the institutional racism of bureaucracies that deter, suspect, stigmatize, check, and interrogate them. It is precisely these institutional arrangements that cause Black people to come into contact with the criminal justice agencies, whether as victims, suspects, or defendants.

The legal and ideological construction of Black people as excluded from subjective citizenship, conditions the social and official responses to them. Therefore, immigration and justice officials operationalize legal definitions of citizenship, which effectively renders all Black people as questionable citizens. Concomitantly, Black people are represented as unwelcome, unwanted, and dishonest in popular and political discourse. Within such historical, sociopolitical, and legal definitions of citizenship, it is hardly surprising that agencies of regulation, whether immigration officials or criminal justice agencies (particularly the police), continue to regard Black people as problematic by definition. Both their status and their actions are regarded as suspect. This result has profound consequences for criminal and social justice.

To possess citizenship is to be a full member of the community and to enjoy the civil, political, and social rights that constitute membership. Having examined the promise and the actuality of citizenship for Black people in contemporary Canada, it has become clear that the exclusion from legal and respectable citizenship, from full participation in social and political life, exacerbates their legal status and their actions remain questioned and regarded as suspicious. This leaves, further, a space for criminalization. Criminalization *per se* is an important feature of the discourse of exclusion. As Gilroy (1987) reminds us, Black lawbreaking supplies the historic proof that Blacks are incompatible with the standard of decency and civilization which the nation requires of its citizenry. Thus, when Black people break the law, they are seen to close a rhetorical circle which links race and non-citizenship with criminal activity. For many Black people in Canada, the promise of full citizenship has not materialized, but the experience of racism and exclusion continue to be all too real.

Notes

1. On June 15, 1995, new immigration legislation tabled by then-Minister of Citizenship and Immigration Sergio Marchi received Royal Assent in the House of Commons. The legislation, Bill C-44, included several important amendments to *Canada's Immigration Act R.S.C. 1985 c. 1-2* with respect to the rights of permanent residents, Convention Refugees, and sponsored applicants for immigration deemed to be a "danger to the public" by the

Minister of Citizenship and Immigration. The Immigration Minister has long had the authority to issue a security certificate against a permanent resident or other individual in Canada, after having initiated procedures with the Canadian Security Intelligence Service and in the Federal Court on an *ex parte* basis. Upon issuance of a security certificate against an individual, the individual is rendered inadmissible and liable to deportation from Canada, without the right of appeal to the Immigration and Refugee Board (Appeal Division) (s. 19; 27; 39-40 of the Act). The Act has also vested the Immigration and Refugee Board (Adjudication Division) with the power of detention where individuals in Canada may pose a danger to the public (s. 103(3)). Bill C-44, an Act to amend the Immigration Act and the Citizenship Act and to make consequential amendment to the Custom Act, S.C. 1995, c.15, expanded existing provisions in the Act relating to the Immigration with broad powers to issue an opinion respecting the branches of "dangerousness" of individuals, without consolation or concurrence from other adjudicative branches of government such as the Immigration and Refugee Board of Federal Court, the threshold offense carrying a possible maximum penalty of ten years or more. Additionally, permanent residents, returning residents, visa holders, and Convention Refugees appealing removal orders lose their right of appeal if found to constitute a danger in the opinion of the Minister's delegate. (See "The Implementation and Impact of the Danger to the Public Provisions Of Subsection 70(5) of the Immigration Act of Canada." A Discussion Paper. Toronto.

2. The *Operations Memorandum EC 95-05*, issued by Immigration Canada on July 17, 1995, sets out guidelines relating to the profile considerations and types of offence which might give rise to a referral for a dangerous opinion. The provisions would normally apply to persons whose offenses involved violence, narcotics, trafficking, sexual abuse, or the use of weapons, offenses that are punishable by a term of imprisonment of ten years or more (*C-44 Enforcement Procedures, Operations Memorandum 95-05*, Citizenship and Immigration Canada, National Headquarters, 17 July 1995)

3. "Toronto's homicide rate, at two in 1000,000 has remained steady for years...Does murder become an issue only when it moves uptown? Guns and Roses (Geogina Leimonis murder)." *Toronto Life*, 28(9), June 1994, p. 7.

4. "The government is responding to a perception that there is a wave of crime and that immigrants cause it. That's patently false, but the government deals in political perceptions." Toronto Immigration lawyer, cited in "Critics say Law Makers Scapegoat of Immigrant Criminals." *Canadian Press Newswire*. September 1996.

5. "Manhunt Police have nabbed 3,600 Alien Fugitives, with 9,000 on the Lam." *Maclean's* (Toronto- Edition), 109(33), 12 August 1996.

6. "Immigrants with criminality have become a political football... Politicians want to be seen to be doing something and this is doing something. Whether it's the right thing or not is another matter." Gordon Maynard, Canadian Bar Association. Immigration Law Section, Canadian Bar Association 1996 Annual Meeting, cited in "Lawyers Condemn Process for Deporting Dangerous Criminals." *Canadian Press Newswire*, 27 August 1996.

7. The following are examples of quotes from newspaper stories that illustrate the media's role. "Tuesday April 5. Three men stroll into the Davenport just dessert...randomly gun down patron Georgina Leimonis. Then outrage. Cries go out for tougher gun control legislation, stiffer immigration restrictions." "Guns and Roses (Georgina Leimonis Murder)." *Toronto Life*, 28(9), June 1994, p. 7.

8. The Toronto Sun ran a frenzied campaign that practically incited lynching...a photo of a Jamaican-born suspect was headlined "Hang Him." It became clear that this "debate" in the Sun was not about immigration it was about race. Toronto Sun billboards around the city displayed large mug shots of fictitious criminals...with "Deported" stamped on their faces and the paper's slogan "We'll be there" (Arif Noorani and Cynthia Wright, *This Magazine*, December 1994).

9. I use the expression "communities" deliberately to indicate that the African/Afro- Caribbean segment of the population is/was not only divided along class lines but also contained other internal fractures such as gender, religion, and nationality.

10. "New Laws Tighten Immigration Lobbying Rules (Bill C-44)." *Canadian-Press-Newswire*, 16 June 1995.

11. See Kenneth B. Nunn, "Law as a Eurocentric Enterprise." *Law and Inequality: A Journal of Theory and Practice* xv(2), 1997. Richard Delgado and Jean Stephancic, *Critical Race Theory* and David Kairys ed. 1982, *The Politics of Law*.

12. I rely significantly on the work of Kenneth B. Nunn here, in the usage of this term. He adopts a position that law is "Eurocentric"...because it expresses attributes that are characteristic of European culture. Law in this sense, would be "Eurocentric to the extent it was a tool of prejudice or bias....it may be used to describe the practice of viewing history."

Chapter 5

The Police in Community Policing: A Black Viewpoint

As the most visible arm of the criminal justice system, the police elicit some emotional reaction from practically every component of Canadian society. To many, the police are seen as guardians of justice, protectors of life, property, and "the Canadian way." To others, especially the poor and large segments of the minority communities, the police are viewed as oppressive and disruptive forces of control to be avoided at all costs. To a great extent, these differences in attitudes regarding the police stem from the differing cultural, political, ideological, and environmental backgrounds of our Canadian population. This chapter looks at the function of police in community policing. What follows is not meant as a banal indictment against the police. On the contrary, it is intended to critically analyze the function and role of the police system in community policing.

The Police and the Public

Generally, Canadians hold their police officers and police forces in high esteem (Seagrave 1997). An understanding of the extent and nature of this support is important insofar as it provides insights into why police-community relations policies succeed or fail and help in comprehending the background to police initiatives. In a nation wide survey, Normandeau and Leighton (1990) found that Canadians were most supportive of their police forces, with nine out of ten satisfied by the performance of their police officers. This overall support was also consistent with the regional findings of Murphy and Clairmont (1990) who showed that 80 percent of the residents of Nova Scotia described police-community relations as "excellent." In Ontario, these figures were also mirrored by Yarmey (1991). Overall, Canadians have positive perceptions of the police, particularly on measures of approachability and enforcing the law (Statistics Canada 1991).

One of the most visible forms of control is exhibited through the police, and the level of overall support is high as indicated by the above studies. It is important to recognize that certain groups in society are more supportive of the police than others. As pointed out by Jones (1977), "we could expect those gaining privileges from the results of police activity to respond favorably toward such an institution" (p.26). On the other hand, those who view themselves as victims, or who *are* victims of the system, tend to be somewhat hostile toward the police. They are alienated in that they have been excluded in a variety of ways (as argued in previous chapters of this book) from the mainstream of life. They are treated as outsiders or intruders in Canadian society. To them, the police function to maintain the status quo for the White privileged elites. It would, therefore, appear that while certain groups within Canadian society do not hold positive views about the police, overall the police receive considerable public support.

The Structural "Changes"

Few people would deny that the shape of policing has changed dramatically, if not fundamentally, over the past two decades. The structural changes began in the late 1960s and early 1970s, in particular with the development of reactive "fire-brigade" policing, the emergence of a quasi-military "third force" concealed inside the ordinary police, and the expansion of surveillance in terms both of technology and the number of subjects. Ericson and Haggerty (1997) and Scraton (1985) show us that as society has become more fragmented, the focus of police work has shifted from traditional modes of crime control and order maintenance to the provision of security through surveillance technologies designed to identify, predict, and manage risk. This form of consolidation grew quickly in the aftermath of urban unrest of the 1980s and 1990s in Britain, the United States, and Canada. In Canada during the spring of 1990, it was all too clear in the policing of the Oka crisis (Fleming 1994) and the policing of Black people through the Immigration and Citizen Act that policing was problematic. At the same time, formal police powers have been considerably enlarged, most notably expanding into other domains, like the education system in Ontario.

Any critical definition of "community policing" defies simplistic approaches. At times, it seems that there are as many definitions as there are people talking about it, from Police Chiefs who claim that they are doing it, to others who dismiss it as public relations. What it entails in practice is virtually everything from putting officers back on foot patrols through programs of community relations, juvenile liaison, community involvement, the all-embracing theory of John Alderson, to the "multi-agency" or "corporate" approach to policing developed by Sir Kenneth Newman in 1982. What all these approaches and practices have in common is that they involve attempts by the police to deal with people whose support of the police appears (to the police) to be weak or non-existent, and which therefore requires strengthening, harnessing, or even creating.

The ideological construction of the involvement of Blacks in crime, in particular street crime, provided the basis for developing strategies of control aimed at keeping young Blacks off the streets and keeping the police in control of particular areas that have become identified both in popular and official discourses as crime-prone or potential "trouble spots." It is necessary to examine or re-examine briefly the notion of the consensual basis of Canadian policing, a basis to which much community policing claims to be returning.

The Myth of Consent

It is the accepted wisdom that the consent of the public lies at the heart of the Canadian policing tradition and much of the writing on community policing harkens back to the golden age of policing when the police could count on the active support and consent of the Canadian public. It is doubtful whether such a golden age ever existed, as a growing body of historical work shows, for example the *Cole-Gittens Report* (1995).

In addressing this point I turn to Robert Storch (1981). He points out, "the advent of the new, professional police in the mid-nineteenth century has shown that the imposition of the modern police was widely opposed, often violently, as the police came as unwelcome spectators into the very nexus of urban neighbourhood life" (p.75). Such resistance continued into the twenty-first century and although its forms have changed from anti-police riots, its content differs little. The disenchantment with the police function has led governments to study the problem. These studies and subsequent calls for reform are the result of the Canadian people being historically torn between a desire for maximum personal freedom and a need for order. In effect, the police and the government had recognized this for sometime,[1] however reluctantly they might acknowledge it in public. Community policing work with the Black communities (as developed in chapter 3) existed as a separate and distinct concept of police work. This liaison attempted to deal with the lack of consent and support of various sectors of society. Lawrence Roach (1978), a British police historian on the subject, states "every policeman was a community relations officer, seeking...to retain and reinforce the goodwill of the public towards their police force" (p.20). What brought about this development was not just post-war changes in attitudes toward authority (now seen as something to be questioned rather than respected), nor increased social mobility within society as some authors and government reports may claim, but an increased mobility between cultures. Interestingly, it was this, according to Roach (1978), that "had broken down the homogeneity of the community to the extent that policemen were no longer able to clearly identify with the people they sought to serve" (p.25). However, attention must be drawn to the institutionalization of culture under the Federal Multicultural Act in Canada. This Act set the tone and a point of reference for most discourses as waves of immigrants came to live in Canadian society

    This development of police-community relations work took place at the same time as the Canadian state had (re)institutionalized racism in its immigration controls and increasingly intervened in race relations through anti-discrimination laws and community relations organizations (following the pattern of Britain and United States). The focus on police-community relations was an attempt to manage the Black population and mediate its opposition and distrust. However, in looking for the reason for this distrust, the police and official discourse (government inquiry) located it in the cultural background and social structures of Black people (for example, the ways in which some Black communities carry on their lives, which may not fit within the dominant "order," such as late night parties with loud music), which often prompted legal or authoritarian responses. It is clear that the development of a "specialist" concept of community policing work absolved the police institution in general of any notion that they were accountable to the Black community. Not surprisingly, community policing to date hardly touched the surface of ordinary, everyday racist policing such as racial profiling.

As the history of police-Black relations in Toronto shows, police-community relations work did nothing to elicit the trust of Canada's Black population. Rather, police-community relations engaged in practices that affirmed the distrust and, at the same time, has placed Black communities apart from the rest of society, a "group" requiring its own special liaison measures. As a result, these practices were, and continue to be, a form of exclusion.

Involvement in the Community[2]

Police involvement in communities, either informally through beat policing or more formally through urban programs and other inner-city projects,[3] places them in a powerful position. They have access to various areas, communities, and information that would otherwise not be available to them. They often control money and the allocation of resources and, thus, they inevitably come into close contact with other agencies. This inter-agency relationship is never one of equality. The police always emphasize that they are in a unique position to provide leadership, initiative, and generally act as a focal point for joint work. In other words, they are in a position to determine priorities, to control the direction of activities, and to isolate and marginalized those who disagree or criticize.

At this point, I will digress slightly from the central focus to show how information/knowledge presented by the police serves as a tool to marginalize. Here, I turn to *Inscription Devices*.

Inscription Devices

The notion of inscription is an important and often relied upon tool in analyses of government (Miller and Rose 1990; Rose and Miller 1992) and ruling (Smith 1988; Smith 1990). I am suggesting that inscription devices assume an important position in the constitution and re-constitution of "community" and that "community" is an object of knowledge through a detailed analysis of such mechanisms.

A useful definition of inscription is provided by Smith (1988): "'Inscription' is used here to point out the practices involved in producing an event or an object in documentary form as a 'fact' about the world" (p.171). Inscription devices include, among others, crime statistics, population censuses, flow charts, surveys, and written reports such as police reports or the minutes of committee meetings. Via these techniques, Rose and Miller (1992) argue that "reality is made stable, mobile, comparable, combinable...It is rendered in a form in which it can be debated and diagnosed" (p.174). Inscription mechanisms are thus types of knowledge, which enable the governance of various objects.

However, as Miller and Rose (1990) observe, inscription is a form of governance in itself. To illustrate, lets consider a hypothetical questionnaire. Consider a questionnaire distributed to community police committees that ask respondents to indicate the issues that had been examined and responded to by these groups. Committee members are to check off any or all applicable categories. The available choices are: crime/drugs/prostitution; traffic/parking; youth; and "other." In filling out this survey, committee members are urged to think of these issues as distinct from each other. Drugs, for example, are separated from parking. In other areas, and for other purposes, they are frequently paired together as evidence in the campaigns by police against drugs in the Jane/Finch area. Also, the linking together of crime, drugs, and prostitution encourages respondents to regard prostitution as a crime although legally it is not. The choices available in such a survey are significant to the authors as opposed to the respondents. The issue of race or race relations does not appear, although most of the community organizations at the time was formed on the basis of bad race relations between the police and the Black communities. Such a questionnaire further solidifies this point of inscription. As a form of governance, techniques of inscription seek to operationalize the desires of authorities. People are linked to various authorities by and through, inscription devices such as surveys, questionnaires and reports. This is an illustration of how police attempts to control and reduce levels of criticism against them. Thus, the mechanism of inscription is a means of governing at a distance.

The Police Survey: A "Community" Profile

The survey was distributed in the winter of 1990 by 31 Division's community patrol officers to homes, apartment buildings, and businesses within the Jane/Finch area. The questionnaire was designed to elicit the opinions of "community members" on policing issues within this locale in conjunction with the establishment of the 31 Division Police Citizen's Committee. There is scant information on this survey and even less on its results. Community policing Staff Sergeants, with whom I spoke informally, did not have any knowledge of the location of the survey nor its results; nor were any of the committee members interviewed. Regardless, it is apparent that community-policing officers do rely on data derived from this survey when accomplishing their routine tasks.

Crime rates, for example, are utilized in the performance and evaluation of policing duties. At committee meetings, generally, a report is given by the police regarding the nature and status (e.g., solved or unsolved) of crimes that have occurred in the Jane/Finch "community." In fact, such a report by the community policing patrol is part of the standard agenda. This report often contains references to increasing or decreasing rates of crime. It seems plausible to suggest that the base line for such commentary or comparison is the crime rate that was part of a larger police survey package. Indeed, prior to the survey, statistical analyses of crime would not have been confined to, nor limited by, the geographical parameters of the Jane/Finch area. Such analyses would have yielded information about crimes that occurred within the boundaries of 31 Division, an area much larger than the locale in question.

Despite the scarcity of (formal) information on the survey, it can be safely assumed that it is standard procedure or protocol for the police prior to, and during, the initiation of a community-based policing project. Evidence to substantiate this claim is derived from the Ministry of the Solicitor General document *Profiling the Community* (1990b). This report advocates the compilation of a statistical profile of the "community" to be policed. Data are compiled to ensure that policing proceeds in an efficient and effective manner in accordance with the desires and needs of the community. Since the community's needs and desires are subject to change, the profile must be continually updated.

Both the police survey and the community profile of which it is a part are inscription devices. As such, they generate knowledge of the community in a way that makes that knowledge stable, combinable, and mobile. Information is amassed that can be used later on in intervention efforts, as previously indicated in the discussion on crime rates.

The police survey and the community profile enable future intervention efforts; they are also a way of governing at a distance. For example, the aforementioned Ministry of the Solicitor General document (1990b) offers the following as possible sources of knowledge for a community profile: Neighborhood associations; Statistics Canada; social service agencies; city/municipal planning departments; business groups; government departments; investigations; community leaders; citizen advocacy groups; schools; police officers; other law enforcement agencies; police research departments; and community surveys. Most of these institutions are not tied to the locale on which the knowledge is to be provided. The community, it appears, is constituted "extra-locally" (Smith 1987). As Ericson (1992) noted, "community" is constituted by institutions that are not constrained by local or geographical boundaries. The concept of the community is not only elusive but fictive as operationalized by the police.

Community Policing: A Taste of Aldersonism

The inter-agency cooperation between the police and other agencies that developed over the years was either confined to a particular section of society, such as young people under the Young Offenders Act, or else limited to one particular geographical part of a police service area and, more often than not, was regarded as "experimental."

John Alderson's (1979) work *Policing Freedom* was, and continues to be, influential and important, both because of the practical and theoretical legacy left with other officers and his advocacy for community policing in political quarters. *Policing Freedom* was a response to what he saw as the era of the "technological cops who rarely meet their public outside conflict or crisis" (Gordon 1987:128). He advocated "pro-active" policing, as distinct from policing which is reactive—that is, merely responding to events—or even policing that is merely preventive. Pro-active policing has all the elements of preventive policing but goes beyond: "setting out to penetrate the community in a multitude of ways...to reinforce social discipline and mutual trust" (p.132). This signifies that a high level of coordination and cooperation at the various levels of the government is needed. It is with these principals that the Metropolitan Toronto Police Services embarked on its journey toward implementing a style of Community Policing.

Black Men and the Police

As a group, Black men have an endless supply of police harassment stories. These include being mistaken for a criminal, being treated like a criminal, being publicly humiliated, and, in some instances, being called derogatory names. Often their encounters with the police arise from being stopped in their cars. They are subject to vehicle stops for a variety of reasons, some legal some not, including:

- Driving a luxury automobile (e.g., BMW, Lexus, Mercedes, etc.)
- Driving an old car
- Driving in a car with a White woman
- Driving in a car with other Black men
- Driving early in the morning
- Driving late at night
- Driving a rented car
- Driving too fast
- Driving too slow
- Driving in a low-income neighborhood, known for its drug traffic
- Driving in a White neighborhood
- Driving in a neighborhood where there have been recent break-ins/burglaries
- Fitting a drug courier profile

It seems that no matter what Black men do in their cars, they are targets for criminal suspicion. It is so commonplace for Black men to be pulled over in their vehicles that this practice has acquired its own acronym: "DWB" (Driving While Black) (Gates Jr. 1995; Harris 1997; and Russell 1998).

Police harassment comes in many forms (see the *Office of the Police Complaints Commissioner Annual Report* 1995). It is also demonstrated by the number of times Black men are stopped, questioned, and assaulted by police as they go about their daily lives. There are clear distinctions between police harassment and police brutality. According to Kappeler, Sluder, and Alpert (1994), police brutality "typically refers to the unlawful use of excessive force" (p.23). For Black men, consistently negative encounters with the police have blurred the line between harassment and brutality. For Black men, who are more likely to be stopped by the police than anyone else, each stop has the potential for police brutality.

The remarks made by the young Black men who participated in focus groups, as detailed in chapter 2, attest to the general fear that Black men have of the police. Many have developed protective mechanisms to either avoid vehicle stops by police or to minimize the potential for harm during these stops. The primary shield they use is an altered public persona. This includes a range of adaptive behaviors (for example, sitting erect while driving, traveling at the precise posted speed limit, avoiding certain neighborhoods, not wearing certain head gear, and if they do wear a baseball cap, wearing it with the peak to the front). Black men are used to structuring their encounters with police during car stops: placing both hands on the steering wheel, responding to an officer's questions with "sir" or "ma'am," and keeping the car radio down before being told to do so. Some Black men are wise to take measures like these because studies consistently show that a suspect's demeanor influences whether he will be arrested or not.

The difference in experiences with law enforcement may explain why Blacks, Whites and others races have contrasting impressions or beliefs about the legitimacy and trustworthiness of the police's treatment of Blacks. However, there is a sense in which all such accounts as explained in this book can be considered counternarratives, or fragments of what I call "subaltern knowledge." This is knowledge that disputes the tenets of official culture. They do not receive the imprimatur of editorialists or of network broadcasters, and if they do, they are not seriously entertained. When they do surface, they are given consideration primarily for their ethnographic value. It is an official culture that treats their (Blacks) claims as it does those of Marxist deconstructionists in the academy; these phenomena are treated as things to be diagnosed, deciphered, given meaning—that is, *another* meaning.

Citizens who do not face the daily threat of being detained largely because of their race are unlikely to understand how burdensome these stops can be. To someone who is pulled over by the police once a month for no apparent reason other than his race, the stops become painful experiences. Race-based policies, Russell (1998) tells us, "pit law enforcement against minorities and create an unbreakable cycle" (p.88). This, in turn, "generates statistically disparate arrest patterns, which in turn form the basis for further police selectivity by race" (Jones 1977:36). Again, as the focus group attested "what many Whites view as the police 'doing their job' is viewed by many Blacks as harassment." Thus, beyond causing harm to Black men, race-based police stops are also harmful to the larger society. There is a societal cost in perpetuating inaccurate stereotypes, which produces exaggerated levels of fear and more pronounced levels of scapegoating such as racial hoaxes (we all know too well the Susan Smith's case of 1994). See Russell (1998) use of this case.

Many Blacks believe that their anti-police sentiments are justified by the racially discriminatory practices of the police. Particularly for young Black men, the police represent public enemy number one. As Russell (1998) warns us, "giving short shrift to the problem of excessive targeting of Black men hampers our efforts to reduce crime" (p.46). For example, the perception that the police unfairly target Black men may make some Black judges and jurors less likely to believe police testimony (*Cole-Gittens* Report 1995; Worden and Shepard 1996). The spillover effect may also make some Blacks less likely to report crime and others less likely to cooperate with police investigations. The research I have conducted reveals that the issue of police abuse is downplayed in Canada because national data are not available.

The Shape of Things: A Discussion

If the police and the Canadian criminal justice system function to maintain the existing societal arrangements, we can expect those uncomfortable with the current arrangement to express their dissatisfaction. Blacks' encounters with the police, the most visible arm of White authority as some may argue, is of a negative nature. Unfortunately, the problem is much larger than White police offices and Black communities. The White police exist and practice their "art" at the pleasure of a very complex, diffuse, and economic power structure that benefits from the existing arrangements.

Some will argue, as did the *Clare Lewis Reoprt and Cole-Gittens Report*, that the addition of Black police officers will create better working relationships between police and Black communities. As a matter of public policy, many agencies including the police have adopted such strategies. Fanon (1963) refers to such strategies as neo-colonialism, in that they effectively involve the oppressed in the very system that oppresses them. Another correlate is that employing Blacks as police officers and in other "civil service" capacities successfully co-opts the most able away from efforts directed toward the collective benefit of Black communities. Actually, there is a residual benefit to White society in this type of arrangement: Black "civil servants" provide other Blacks with role models that are beneficial to maintaining existing social arrangements. As things presently exist, in order to be considered for the middle-class life of a "civil servant," one must be a law-abiding citizen who at least outwardly accepts the social customs of Canadian society.

While some deviation is accepted, a drift too far to the right or left virtually eliminates the chance for public employment. So, to a large extent, every Black hired in "the system" pays a double dividend to White society. She/he immediately helps the system function more smoothly and serves as a role model for untold numbers to follow. While well-intentioned Whites who hire Blacks and Blacks who are hired will have difficulty with these assumptions, it is evident that such hiring practices do result in such an arrangement.

This paradox presents a serious problem to politically conscious people of all colors who are concerned about social justice in society. It suggests that they must move beyond existing assumptions that more Blacks in the service of the criminal justice system will remedy existing problems. Inclusion or exclusion may be only part of a temporary solution. The real issue, or issues, may be in developing something entirely new, something that more clearly addresses the major institutions that cause the problems. Indeed, community control of the police could go a long way in reducing Black alienation from the police and the criminal justice system. Law enforcement officials generally oppose community control of the police on the basis that it is inefficient. They maintain that police autonomy maximizes efficiency and reduces the possibility of political influence. Blacks contend that this is only disguising the issue of police racism. However, if we locate police services as complex social organizations, in general, that seek to perpetuate themselves by continually justifying their existence, then we secure a sense of another aspect of the police in community policing.

The problem is not that community policing is not an alternative to reactive policing. More important, community policing is an attempt at the surveillance and control of communities by the police, an attempt which operates under the guise of police offering advice and assistance, and which is all the more dangerous because it not only merges the activities of different agencies of the state, but does so under the control and direction of the police. According to Lee Bridges (1985),

> community policing merges at the local level the coercive and consensual functions of government, enabling the police to wield a frightening mixture of repressive powers, on the one hand, and programmes of social intervention, on the other, as mutually reinforcing tools in their efforts to control and contain the political struggles of the black and working class communities. (P.80)

This sums up quite well how community policing offers no prospect of greater democratic control of the police and policing. Indeed, community policing has come to the fore precisely at the same time as there has been widespread demands for greater public accountability and control of the police.[4] As I have argued throughout this book, whatever *form* community policing takes, true control of community policing remains firmly in the hands of the police.

Instead, we must locate community policing in the context of the increasing disciplining of society by the state. While its implementation and impact have been modest and, indeed, there is a certain deconstructionism going on in police research, there are signs too of resurgence, a resurgence linked to developments in the police community more than academic support and rooted in an intrusive, aggressive policing style. With the latter focusing symbolically on quality of life issues, it may represent a success for conservative ideologists and be consistent with a Foucauldian model of state expansion and control.

Such discipline takes many forms (for example, the use of the family to control children and keep women in the home, the criminalization of whole sections of society, urban programs, and, of course, direct repression by the police using new technologies and unprecedented new powers). Community policing recognizes that such open control may be counterproductive and may seek to penetrate communities to break down community resistance, to engineer consent and support for the police, and to reinforce social discipline. It is an aspect of what Stan Cohen (1985), writing of Foucault's "punitive city," has described as a "correctional continuum" that involves the proliferation of agencies and services, finely calibrated in terms of degree of coerciveness or intention or unpleasantness, which points to a future when it will be impossible to determine who exactly is enmeshed in the social control system and, hence, subject to its jurisdiction and surveillance at any one time.

Community policing is but one aspect of this continuum of discipline and is all the more dangerous because it appears to offer an alternative to unwelcome police practices and strategies and, at the same time, a promise toward reducing crime by building community support thereby improving the social conditions of the inhabitants of Ontario's inner cities. Therefore, community-policing (usually quite undefined) gains support in all parts of the political spectrum, between classes and among various races, while critical accounts are dismissed as pessimistic or unrealistic.

Notes

1. See Clayton Mosher's 1994 draft working paper, "Crime and Colour, Cops and Courts: Systemic Racism in the Ontario Criminal Justice System: In Social and Historical Context; 1892-1961." It should be noted that within the *Cole-Gittens Report* no reference was made to its findings.

2. Generally envisioned as the most effective approach to contemporary policing, community policing provides much of the framework for discussions of the future of policing in Canadian and international circles (see in Canada, Normandeau and Leighton 1990; in the United States, Trojanowicz and Carter 1988; and in Britain, Leon 1989). Despite this widespread consensus, various authors have expressed concern about the implementa-

tion, viability, and impact of community policing projects (See Bailey 1989; Mastrofski 1988).

3. *Council on Race Relations* and *Policing, Race Relations and Policing Unit*—both were funded by the Ministry of the Solicitor General and Correctional Services and are now out of existence due to lack of funding. Those that are now in existence are the African Canadian legal Clinic, Toronto's Task Force on Community Safety, and the Chinatown Community Police Liaison Community, to name a few.

4. The Special Investigations Unit was formed in April 1992. The Mandate of the SIU is to investigate in circumstances of serious injury to or death of a member of the public caused by a police officer (see George W. Adams, Q.C. 1998, Consultation Report Concerning Police Cooperation with the Special Investigations Unit; See also Bill 105, the Ontario Police Services Amendments Act that folded the Ontario Civilian Commission on Police Services, the Police Complaints Commissioner and the provincial Board of Inquiry into The Police Conduct Commission).

Chapter 6

Afterwards

That few Canadian academics are sought out to provide insight into issues of police/Black relations, particularly police shootings, reflects the general public's estimation of the intellectual class's ability to provide answers to the serious problems of race in Canada. Part of the reason for such a perception may lie in the priorities that guide the production of knowledge and the definition of theory. The breadth and complexity of the Canadian policing system and its history has only been approached by a few in academia. Some studies celebrate the lack of informed critical or race debate. There are, however, some important studies dealing with an aspect or a single force within the system (Ericson 1982; Normandeau and Leighton 1990a, 1990b, 1991; Rosenbaum 1994). The lack of academic contributions more generally would seem to have been partly remedied by a number of Commissions of Inquiry and Task Forces into relations with Black communities in Canada. In spite of all this, it seems that policing is increasingly being forced toward the forefront of the political agenda in Canada. Especially questions through the community regarding the role of the police in a changing society. Within this context, the shortcomings of race-relations theory can be separated into at least two categories.

Beyond the Black/White Dichotomy

Recent books on race, such as Andrew Hacker's *Two Nations: Black and White, Separate, Hostile, Unequal* helps to embed, further, the construction of race relations as a binary opposition. Although the theoretical framing of race relations in Black/White terms has substantial historical and contemporary grounding, the shooting of Edmund Yu reveals that such essentialism misses many of the factual complexities in contemporary urban politics. The treatment of Asian Canadians in the media may reflect more about relations between White and Black Canadians than about relations between Blacks and Asians. The embrace of the model-minority myth by the media becomes a bear hug, particularly at times when Black/White tensions intensify and White Canadians wish to discipline Black communities. The fact is that Asians are the fastest growing immigrant group in Toronto and the oppositional Black/White character of the race-relations debate excludes discussions of the colors in the middle, now inexorable parts of the Black/White spectrum.

White liberal or progressive guilt has been focused largely on the historic exploitation of Blacks by Whites. While there are important structural and historical bases for this concentration, the contemporary realities and demographics of racial groups in Canada necessitate a broader discussion. Undoubtedly, many scholar/activists' direct involvement with the civil rights movement has been confined to struggles waged by Black communities. However, uncritical acceptance of the dichotomous Black/White character of race relations by such scholars obscures the role of Asians, Native Canadians, and people of Indian descent to the detriment of a more differentiated understanding of contemporary race relations, racism, and the struggles to end racial oppression.

Typically, non-African Canadians/American people of color are categorized as either Black or White if they are discussed at all. Asian and people of Indian descent are often summarily included with Blacks under the "people of color" rubric or sometimes referred to as the "Other." The ubiquitous internalization of the model-minority myth by the general population and academics leads to the invisibility of Asian Canadians in the racial landscape. An awkward silence has descended upon liberal and progressive circles analyzing events such as the shooting of Edmund Yu. Many publications avoided discussing Asian Canadians altogether, thereby sidestepping the troubling interracial conflicts among Asian Canadians and Black Canadians. Some boldly categorized Asian Canadians with a contempt usually reserved for the dominant majority, characterizing the immigrant shopkeepers as a primary antagonist of Blacks. Sadly, the 'neo-conservatives' embrace of Asian Torontonians, combined with liberal and progressive neglect or contempt, may trigger a self-fulfilling prophecy of the model-minority myth. Asian Canadians do not identify with European Canadians and see themselves as very distinct. Yet it is a cold reception they receive from community organizers, coalition-minded politicians, and progressive intellectuals, which excludes them from the people of color organizing and theoretical models.

Even in the so-called "Black/Asian conflict," although Asians are necessarily included in the discussion, the conflict is viewed through the lens of Black/White relationships. In other words, how Asian relationships with Blacks are represented and interpreted often depends upon the latter group's relationship to Whites. Asian Canadians are instrumentalized in a larger public-relations campaign on behalf of Euro-Canadians. Moreover, important class and gender dynamics become obscured by the emphasis on racial differences in the discourse on community policing. The conflict between Asian Canadians and Blacks contains definite cultural differences and racial animosities. But many of the tensions may be class rather than racially based, actually reflecting differences between the store-owning Asian immigrants and Black customers. Violence between shopkeepers and residents exists in inner cities regardless of which racial group owns the majority of stores. The interests of the entrepreneurial class transcend racial differences. The Black storeowner and his/her Asian counterpart stand together on issues of community policing. Both oppose grassroots community efforts to limit police discretion. This is particularly true concerning the relaxation of controls placed on the police related to drug enforcement and the use of the SWAT team in response to home invasions. Often, when the public accepts a reduction of civil liberties because of fear, an increase in police deviance associated with the retraction of those rights is the result.

Scholars of ethnic and racial politics must confront the challenges of racial theory in the twenty-first century. In order to do this, a serious effort must be made to incorporate the histories and the contemporary experiences of people of color between the two poles of Black and White on the racial spectrum, especially those of the new and rapidly expanding immigrant groups. Intellectual activists must grapple openly and critically with the position of each community of color within the complexities of race, ethnicity, class, and gender relations in a post-industrial society. For example, scholars must address the question of class in the Canadian context. What are Asian "mom and pop" store owners—are they petty bourgeoisie? Are they capitalist exploiters? Are they self-exploited? Are they the "owning poor"? Are they middle class? Diversity within ethnic and racial groups must be acknowledged and incorporated into theoretical analyses to avoid essentializing race and obscuring important differences and contradictions in community policing.

Structure, Agency, and Theories for Action and Social Change

Much of the theoretical construction of race relations has employed structural analyses that incorporate a critique of institutional discrimination, historical racism, and modes of economic production. While these factors are important and necessary to understanding the state of contemporary race discourse in Canada, an excessively structural analysis presents those subordinated under oppressive systems as "victims" with little or no recourse. Structuralist social scientists often face problems when attempting to contextually understand their subjects as actors possessing agency.

Scholars of ethnic studies, urban politics, criminology, and race relations can work to build toward a theory for action, a theory for social change. Such a theory would emphasize the experiences and conditions of the oppressed and of those working directly to improve those conditions. In order to do so, intellectuals and activists must know the people and live the experience—the pains, the challenges, and the realities—of racism. They must measure the success of theories by their ability to explain racial problems and to provide solutions to difficult problems. If abstract theories do not prove useful to the folks most affected by community policing, they should be abandoned. A respectful and informed partnership must be created in the Freirian tradition to create more relevant research, pedagogy, and theory to assist those suffering in the affected communities. Intellectual activists must know which leadership is respected and acknowledged in different communities. Further, they must consider how change comes about in different neighborhoods and what relationships exist between structures and individuals. In short, intellectuals must leave their offices and go to where the problems are in order to understand that in which they claim to be "experts."

The roles of political leadership, individual accountability, and community education must be addressed in order to make the transition from rigid, structure-induced victim perspectives to progressive, activist-based perspectives. Some of the worst problems faced by subordinated communities, as I observed during the course of my fieldwork, cannot be resolved or addressed simply by reciting the standard critiques of "the system" or "the man." Serious problems such as drugs, crime, domestic violence, child poverty, homelessness, street people, poverty, and interracial conflict clearly have structural roots. But afflicted communities must seek solutions to the toughest problems here and now, since "the system" will not disappear tomorrow. As one community organizer against substance abuse in Toronto stated, "We cannot afford to avoid problems like drugs and crime by saying these issues will get resolved when we change society."

In this spirit, Asian-Canadian organizers and intellectuals must work with communities to reject prejudices and stereotypes about other people of color that have been adopted from the mainstream culture. Asian-Canadians must address the complaints that too many storeowners are rude and disrespectful to darker-skinned customers, and search for ways to improve relations. The community cannot use the reality of high crime rates that shop owners face to rationalize unacceptable behavior. There must be a better understanding of the fearful, bunker mentality of all shopkeepers, regardless of color. While Asians may not have constructed the international racial hierarchy, they can educate one another concerning its fallacies. Each community, likewise, can do some soul-searching and admit the truths that could produce a stronger foundation for coalition politics and the seizing of Toronto's transformative potential.

Similarly, members of the academic community must critically assess their roles, passive or otherwise, in relation to police shootings in Toronto. Intellectuals of color failed miserably at taking a stand on the shootings of Blacks. Because Asian-Canadian academics failed to speak up and condemn the results rendered in the Edmund Yu shooting, they were complicit in accepting the results of the shooting of Blacks as well. Likewise, Black Canadian scholars could have taken a position on this shooting but failed to do so. Activist scholars must be willing to take a stand on issues and immerse themselves in problem-solving tasks. Specific opportunities to intervene, to help solve conflicts, should be grasped.

A more open-ended, qualitative approach is required to conduct such community-based research, which is far beyond the scope of this book. Many traditional methodologies, which emphasize quantitative methods and analysis in order to posit predictions, are not useful in resolving problems of racism. Politics is not a hard science. Even the hard sciences no longer consider themselves "hard." Scholars of ethnic studies, urban politics, and racism must capture the human, not the mathematical, element in politics. Problem solving must become the focus. Theory must draw from activists and organizers as the generative sources of themes and solutions. If academics cite Foucault, Cornel West, bell hooks, Barley, Manning, Ericson, and Skogan in their discussion on community policing and racism, they should also cite community organizers, such as Dudley Laws, in the Canadian context and privilege his and other community organizers' insights as "expert."

In the analysis of racial consciousness, politics, and policing in the twenty-first century, researchers face new challenges currently unaddressed by both conservative and progressive scholars. A new era recognizing the autonomy and strength of people of color will depend largely on our ability to listen to the voices of the subjects being studied, to position people of color as actors in the research who can provide real insight into the diversity and contingencies constitutive of communities of color. The new scholarship can subvert pervasive perceptions of people of color as either the faceless victims or as romanticized, oppressed revolutionaries carrying out an inevitable historical task and, in so doing, perhaps contribute to a new theory and praxis of empowerment.

Some Implications for Police Organizations

What are the implications of the argument for police organizations adopting community-based policing? If we accept the arguments for community policing as valid, then it would seem that, at the broadest policy level, some self-examination would occur as to whether the police profession should continue to exist in its current form. It is unlikely, of course, that any profession would willingly dismantle itself or even entertain suggestions for radical change. Police organizations and their cultures are highly adaptive, but they resist fundamental change tenaciously. As this book illustrates, the media and other players structure the public's perception of crime and race. Structural and organizational arrangements can be modified to manage public appearances so that the issue of race can continue, even flourish, in the organization. The hiring of Black officers and female officers serves as a useful example here. While this might seem contradictory, police cultures and organizations have a remarkable ability to retain their cultural essence and to continue their activities. A particularly astute officer for a large urban department made the following comments:

> Nothing, though, could have prepared me for the avalanche of crime and police coverage here. After about six months, I began to figure out that crime stories, especially for the broadcast journalists, are the fallback. Crime has been going down over the last several years, but crime reporting is going up, and the public's perception has very nearly mirrored the local news reporting of crime. So I said to the media, "We'll never withhold information from you. We're just going to be more selective about when we provide you sound and video." Some people said that's managing the news and protested. I said, "No, we're just managing the message." (Field notes 1996)

An ethos of autonomy is evident in these remarks that can be analyzed further. As the first line of the criminal justice process, police officers make very authoritative decisions about whom to arrest, when to arrest, and when to use force. To this extent, the police are the "gatekeepers" of the criminal justice system. Police officers cling to their autonomy and the freedom to decide when to use force. The desire for autonomy often exists despite departmental, judicial, or community standards designed to limit the discretion of street enforcement officers (see *Cole-Gittens Report* 1995, chapters 10-11). Personally defined justice, reinforced by subcultural membership, can lead to abuses of discretion, an issue the *Report* did not attempt to deal with. However, any attempt to limit the autonomy of the police is viewed as an attempt to undermine the police authority to control "real" street crime and not as an attempt on the part of citizens to curb police abuses of authority.

Admittedly, contemporary society needs police organizations, but not racism. To that end, we can reconcile the merit that exists in the logic of community policing with the drive to self-perpetuate the police organization. Attention *must* be given to the question of race/ racism. Exactly what do we collectively want the primary focus of our police organizations to be? In a general sense, there would appear to be at least two basic options available.

The first of these options would be to continue with the *status quo* and presume that either the *Cole-Gittens Report* arguments are wrongheaded altogether, or, at least, do not apply to police organizations but to other parts of the criminal justice system. The problem with this approach is that current realities seem to suggest that the *Cole-Gittens Report* did have some insights and some implications for police organizations. On the other hand, the public's perception of safety in society does appear to be declining, especially when the issue of race and crime is the focus of discussions. Concomitantly, police organizations seem to be drowning in a sea of regulations and busy work, neither of which existed to the same extent prior to the institutionalization of community-policing that occurred during this century, particularly in the last thirty years. It is especially sobering to note that community policing is not an isolated phenomenon. The same themes of community ownership and empowerment appear in virtually every segment of Canadian public life.

A second option would be to attempt to halt, or at least slow, the extent to which police organizations focus on community-policing, depending on the geographical location, at the expense of the more essential responsibilities of policing—that is, control or management of crime or delivery of enhanced services that improve the quality of community life and citizens' satisfaction with policing. These latter concerns have been the focus of community-policing programs (Manning 1992) across North America. Within the last decade, police departments across Canada and the United States have adopted community-policing philosophies and strategies in an effort to redefine the ends and means of policing. While community-policing is based upon a concept that police officers, private citizens, local and state agencies can form partnerships and work together in creative ways towards identifying and solving contemporary community problems related to crime, such as fear of crime, social and physical disorder, and neighborhood decay, it has gain little support in some communities. However, in some communities, community policing is seen as one of the means to reduce and prevent crime, and to protect and enhance the quality of life in an urban environment (Trojanowicz and Bucqueroux 1989; Skolnick and Bayley 1989; Goldstein 1990; Wycoff and Skogan 1993; Rosenbaum 1994).

Caution must be exercised. The very meaning of the concepts of "community" and "police" as they are located within the objectives of the political economy of utilitarianism must be re-evaluated. It must be remembered that in the modern era, community itself is constantly disrupted and social life is hierarchically structured, leaving no room for people to create their own community life outside of the civil structures that contain them. The forms of social organization, which takes place in the typical urban spaces, are based not on any sense of "community," which arises spontaneously from the will of its members, but rather develops out of a sense of individualism based on "privacy," and the discourse of "rights." In this sense, people live in so-called communities in which they may not even know their neighbors. It is the position of this book that the sense of community that exists in typical urban neighborhoods comes from people's concern for their property and its protection. It seems that if one conveniently mixes this real-life situation of anonymous social life with abstract notions about supposed communities, one runs into conceptual dangers that seriously problematize attempts at developing a critical stance with respect to issues and forms surrounding policing, community, and the social order. Indeed, Foucault was correct to point out that discussions of community are the modern terrain for the production of "human subjectivity."

If any conclusion may be drawn, it is that community policing as a form of exclusion is a complex, multifaceted, and multidimensional enigma in a racialized society. Thus, there are no simplistic, quick fix, cookbook solutions for the problems of racism. Despite these cautionary notes, a complex, interrelated web of suggestions that focuses on the police institution needs to be presented. At the simplest level, the opportunity structure inherent in the nature of policing presents officers with virtually unlimited chances to engage in racism. Hiring well-qualified and capable officers, providing appropriate training and education on race relations, mandating that supervisors hold officers accountable for their behavior, and centralizing administrative control are all simple means to impede racism in police organizations. Other control mechanisms may be suggested as well. Included among them are the development and endorsement of agency guidelines that are clear, practical, reasonable, and workable. Progressive and consistently applied disciplinary schemes are critical. Internal affairs units, the Special Investigation Unit and review boards, if they are able to function with integrity and handle complaints objectively, can be used as checks and balances to minimize police racism.

Although all of these factors are important in the control of individual police officer racism, they alone cannot ensure the control or elimination of racism in the police occupation. In fact, there are other agencies within the criminal justice system that use many or all of these mechanisms, yet continue to suffer from problems of systemic racism. In order to effect significant change in the police occupation, deeper, more fundamental modifications to the existing social order and the police normative structure are required.

Police racism serves functional roles for society and for police culture. Thus, society has two options. One is to acknowledge the role of racism in maintaining the culture of the police and the existing social order. Although not a remedy for racism, recognition of the functional aspects of police racism is a foundation from which society can begin to understand the realities of both policing and community policing. Public acknowledgment of the parameters of police racism would help to dispel myths such as the rotten-apple theory of misconduct, and portrayals of police as loyal social control agents supportive of the expressed moral order. Unfortunately, this would permit and even encourage racism.

A second, more radical approach would entail modifying the bounds of society's normative system and the police normative structure. Attaining congruity between the two would obviously be an onerous challenge. Public officials would need to refrain from politicizing crime, capitalizing on citizen's fear of crime, imputing a warlike mentality for crime control, linking race and crime, and advancing the misconception that the police can effectively control social disorder and crime itself. The judiciary must send a clear and constant message that police shootings under questionable circumstances will not be tolerated. This then raises the following question: who controls the police who are charged with controlling the dangerous supplement? Ambiguities in the criminal law must be resolved by both the legal and law enforcement communities. This means, among other things, that equitable and ethical approaches to attaining justice must be emphasized, while backdoor approaches often used to circumvent legal requirements must be discontinued. Administrative customs and practices in the police institution must conform to both the stated law and agency policy. In essence, the radical approach suggests restructuring the law, reordering the social structure, and reinventing police culture.

In the final analysis, Canada must confront the most fundamental question of all: How much social change should be made in an attempt to control or eliminate racism? While this book may not provide an answer to the question, we are reminded that any society may have as much crime and racism as it deserves. It remains unclear whether or not it is impossible to control police racism. Without substantial social change, however, it seems clear that attainment of this goal by way of community policing will be improbable at best. Community policing, masquerading as innovation—for example, officers on bicycles in the summer—continues today and is still blurred and undefined in search of a social location. In its search, forms of exclusion emerge which perpetuate racism.

References

Althusser, L. 1971. *Lenin and Philosophy and Other Essays*. New York, NY: Monthly Review Press.
Auletta, K. 1982. *The Underclass*. New York: Random House.
Baker, D., ed. 1994. *Reading Racism and the Criminal Justice System*. Toronto: Canadian Scholars' Press Inc.
_____. 1992. "Race and Class: The Debate Lingers." Unpublished Paper, York University, Toronto.
Balut, J.M. 1992. "The Theory of Cultural Racism." *Antipode* 24:289-99.
Barbalet, J.M. 1988. *Citizenship*. Milton Keynes: Open University Press.
Barrett, M. 1991. *The Politics of Truth – Form Foucault to Feminism*. Stanford: Stanford University Press.
Barker, M. 1981. *The New Racism*. London: Junction Books.
Bayley, D.H. 1988. "Community Policing: A Report from the Devil's Advocate." In *Community Policing: Rhetoric or Reality?*, edited by Jack Green and Stephen D. Mastrofski. New York: Praeger.
Bennett, T. 1992 "Putting Policy into Cultural Studies." In *Cultural Studies*, edited by L. Grossberg, C. Nelson, and P. Treichler. London: Routledge.
Boggs, C. 1976. *Gramsci's Marxism*. London: Pluto Press.
Bridges, L. 1986. "Beyond Accountability: Labour Leadreship and the Rebellions." *Race and Class* 37(4):75-88.
Brown, L. and A. Willis. 1985. "Authoritarianism in British Police Recruits: Importation, Socialisation or Myth?" *Journal of Occupational Psychology* 58:97-108.
Burton, F. and P. Carlen. 1979. *Official Discourse: On Discourse Analysis, Government Publication, Ideology and the State*. London: Routledge and Kegan Paul.
Canadian Press Newswire. 1995. "New Laws Tighten Immigration Lobbying Rules (Bill C-44."
Cain, M. and S. Sadigh. 1982. "Racism, the Police and Community Policing." *Journal of Law and Society* 9(1):87-102.

Calliste, A. 1991. "Canada's Immigration Policy and Domestics from the Caribbean. The Second Domestic Scheme." In *The Social Basis of Law,* edited by E. Comack and S. Brickely. Halifax: Garamond.

Cartwright, T. 1975. *Royal Commissions and Departmental Committees in Britain.* London: Hodder and Stoughton.

Cashmore, E. and E. Mclaughlin, eds. 1991. *Out of Order? Policing Black People.* London: Routledge.

Clarke, M. J. 1987. "Citizenship, Community, and the Management of Crime." *British Journal of Criminology* 27(4).

Cloward, R. and L.Ohlin. 1960. *Delinquency and Opportunity.* New York: Routledge.

Cohen, A. 1985. *The Symbolic Construction of Community.* England: Ellis Horwood Ltd.

Cohen, P. 1979. "Policing the Working Class City." In *Capitalism and the Rule of Law,* edited by B. Fine. London: Hutchinson.

Cohen, S. 1981. "Footprints on the Sand: A Further Report on Criminology and the Sociology of Deviance in Britain." In *Crime and Society: Readings in History and Theory,* edited by M. Fitzgerald, G. Mclennan, and J. Pawson. London: Routledge and Kegan Paul.

———. 1985. *Visions of Social Control.* Oxford: Polity Press.

Cook, D. 1994. "Racism, Citizenship and Exclusion." In *Racism and Criminology,* edited by D. Cook and B. Hudson. London, England: Sage Publications.

Cook, D. and B. Hudson, eds. 1994. *Racism and Criminology.* London, England: Sage Publications.

Ericson, R.V. 1982. *Reproducing Order: A Study of Police Work.* Toronto: University of Toronto Press.

Ericson, R.V. and K.D. Haggerty. 1997. *Policing the Risk Society.* Toronto: University of Toronto Press.

Fleming, T.O. 1994. "The Mohawk-Canada Crisis: Native Peoples, Criminalization and the Justice System." In *Reading Racism and the Criminal Justice System Baker,* edited by D. Baker. Toronto: Canadian Scholars' Press Inc.

Fielding, N. 1988. *Joining Forces.* London: Routledge.

Fitzpatrick, P. 1987. "Racism and the Innocence of Law." Reprinted in *Reading Racism and the Criminal Justice System,* edited by D. Baker, 1994. Toronto: Canadian Scholars' Press Inc.

Foucault, M. 1991. "Governmentality." In *The Foucault Effect: Studies in Governmentality,* edited by Graham Burchell, Cloin Gordon, and Peter Miller. New York: Harvester/Wheatsheaf.

_____. 1979. *The History of Sexuality. Vol.1 An Introduction.* London: Allen Lane.

_____. 1977. *Discipline and Punish.* New York: Pantheon Books.

_____. 1972. *The Archaeology of Knowledge.* Translated by A. Sheridan. New York: Pantheon Books.

Garland, D. 1996. "The Limits of the Sovereign State: Strategies of Crime Control in Contemporary Society." *The British Journal of Criminology* 36(4):445-471.

Gates, H. L. Jr. 1995. "Thirteen Ways of Looking at a Black Man." *The New Yorker*, 23 October, p.56.

George-Abeyie, D. 1990. "The Myth of a Racist Criminal Justice System?" In *Racism, Empiricism and Criminal Justice,* edited by Brian Maclean and Dragan Milovanovic. St. Augustine, FL: The Collective Press.

Gilroy, P. 1982. "Police and Thieves." In *The Empire Strikes Back,* edited by The Center for Contemporary Cultural Studies London: Hutchinson.

_____. 1987a. *There Ain't No Blacks in the Union Jack.* London: Hutchinson.

_____. 1987b. "The Myth of Black Criminality." In *Law, Order and the Authoritarian State*, edited by P. Scraton. Milton Keynes: Open University Press.

Gilroy, P and J. Sim. 1985. "Law, Order and the State of the Left." *Capital and Class* 25:15-55.

Gordon, P. 1987. "Community Policing: Towards the Local Police State?" In *Law, Order and the Authoritarian State*, edited by P. Scraton and J. Sim. Milton Keynes: Open University Press.

Gramsci, A. 1971. *Selections from the Prison Notebooks*, edited by Quintin Hoare and Geoffrey Nowell Smith. New York: International Publishers.

Habermas, J. 1976. *Legitimation Crisis.* London: Heinemann.

Hall, S. 1980. *Drifting into a Law and Order Society.* London: Cobden Trust.

Hall, S., C. Critcher, J. Clarke, T. Jefferson, and B. Roberts. 1978. *Policing the Crisis*. London: Macmillan.

Hall, S., B. Lumley, and C. McLennan. 1978. "Politics and Ideology: Gramsci." In *On Ideology: Center for Contemporary Cultural Studies*. London: Hutchinson and Company.

Harris, D.A. 1997. "'Diving While Black' and all other Traffic Offences: The Supreme Court and Pretextual Traffic Stops." *Journal of Criminal Law and Criminology* 87(Winter):544.

Harris, R. 1992. *Criminal Justice and the Probation Services*. London: Routledge.

Hawkins, D. F. and R. Thomas. 1991. "White Policing of Black Population: A History of Race and Social Control in America." In *Out of Order? Policing Black People,* edited by E. Cashmore and E. McLaughlin. London: Routledge.

Henry, S. 1988. "Can the Hidden Economy be Revolutionary? Towards a Dialectic Analysis of the Relations between Formal and Informal Economies." *Social Justice* 15:29-60.

Henry, S. and D. Milivanovic. 1991. "Constitutive Criminology: The Maturation of Critical Theory Criminology." 29(2):293-315.

Hudson, B. 1994. "Racism and Criminology: Concepts and Controversies" In *Racism and Criminology*. London, England: Sage Publications.

Jackson, P. 1994. "Policing Difference: 'Race' and Crime in Metropolitan Toronto." In *Constructions of Race, Place and Nation,* edited by P. Jackson and J. Penrose. London: UCL. Press, pp 181-200.

James, J. 1996. *Resisting State Violence*. Minnesota: University of Minnesota Press.

Jefferson, T. 1990. *The Case against Paramilitary Policing*. Milton Keynes: Open University Press.

_____. 1988. "Race, Crime and Policing: Empirical, Theoretical and Methodological Issues." *International Journal of Sociology of Law* 16(4):521-41.

Junger, M. 1989. "Ethnic Minorities, Crime and Public Policy." In *Crime and Criminal Policy in Europe,* edited by R. Hood. Oxford: Centre for Criminological Research.

Kappeler, V.E., R.D. Sluder, and G.P. Alpert. 1998. *Forces of Deviance: Understanding the Dark Side of Policing*. Prospect Heights, Illinois: Waveland Press Inc.

References

Kobayashi, E. 1993. "Multiculturalism: Representing a Canadian institution." In *Place/Culture/Representation,* edited by J. Duncan and D. Ley. London and New York: Routledge, pp.205-31.

Lea, J. and J. Young, J. 1984. *What is to be Done about Law and Order?* Harmondsworth, UK: Penguin.

Lears, T.J.J. 1985. "The Concept of Cultural Hegemony: Problems and Possibilities." *American Historical Review* 90(3):567-593.

Lemon, J.T. 1984. "Toronto among North American Cities: A Historical Perspective on the Present." In *Forging a Consensus*, edited by V.L. Russel. Toronto: University of Toronto Press, pp.323-351.

Lewis, C. 1989. *Report of the Race Relations and Policing Task Force.* Toronto: Queen's Printer.

Lister, R. 1990. *The Exclusive Society: Citizenship and the Poor.* London: CPAG.

Maclean's (Toronto edition). 1996. "Manhunt Police have nabbed 3,600 Alien Fugitives, with 9,000 on the Lam." 109(33).

Manning. P. 1988. *Symbolic Communication: Signifying Calls and the Police Response.* Cambridge: The MIT Press.

Marshall, T.H. 1950. *Citizenship and Social Class and Other Essays.* Cambridge: Cambridge University Press.

_____. 1981. *The Right to Welfare and Other Essays*. London: Heinemann.

McIntyre, C.C.L. 1993. *Criminalizing a Race: Free Blacks During Slavery.* Queens, New York: Kayode Publications Ltd.

McNulty, E.W. 1994. "Generating Common Sense Knowledge among Police Officers." *Symbolic Interaction* 17(3): 281-294.

Merton, R. 1964. "Anomie, Anomia and Social Interaction." In *Anomie and Deviant Behaviour,* edited by M. Clinard. New York: The Free Press.

Mensch, E. 1982. "The History of Mainstream Legal Thought." In *The Politics of Law–A Progressive Critique,* edited by D. Kairys. New York: Pantheon Books.

Miller, P. and N. Rose, N. 1990. "Political Rationalities and Technologies of Government." In *Text, Contexts, Concepts: Studies on Politics and Power in Language*, edited by S. Hanninen and K. Palonen. Finland: Finnish Political Science Association.

Mills, C.W. 1988. "Getting out of the Cave: Tensions between Democracy and Elitism in Marx's Theory of Cognitive Liberation." Presented at the Annual conference of the Caribbean Studies Association, May 25-27, Guadaloupe.

Murphy, C. 1988. "Community Problems, Problem Communities, and Community Policing in Toronto." *Journal of Research in Crime And Delinquency* 25(4):392-410.

Murry, C. 1984. *Losing Ground.* New York: Basic Books.

Noorani, R. and C. Wright. 1994. "Hang Him." *This Magazine,* Toronto.

Normandeau, A. and B. Leighton. 1990. "A Vision of the Future of Policing in Canada: Police Challenge 2000." Background Document, Ottawa: Police and Security Branch, Ministry Secretariat, Solicitor General Canada: Queen's Printer.

Nunn, K.B. 1997. "Law as a Eurocentric Enterprise." *Law and Inequality: A Journal of Theory and Practice* 15(2):323-371.

Omi, M. and H. Winant. 1987. *Racial Formation in the United States: From the Sixties to the Eighties.* London: Routledge.

Parekh, B. 1974. *Jeremy Bentham.* London: Frank Cass.

Pratt, M. 1980. *Mugging as a Social Problem.* London: Routledge and Kegan Paul.

Reiner, R. 1985. *The Politics of the Police.* Brighton: Wheatsheaf.

_____. 1988. "British Criminology and the State." *British Journal of Criminology* 28(2):268-88.

_____. 1989. "Race and Criminal Justice." *New Community* 16(1):5-22.

Robinson, V. 1993. *The International Refugee Crisis and Canadian Refugee Policy to 1980: A Historical Overview.* Great Britain: Anthony Rowe Ltd.

Rock, P. 1988. "The Present State of Criminology in Britian." *British Journal of Criminology* 28(2):188-99.

Rose, N. 1993. "Government, Authority and Expertise in Advanced Liberalism." *Economy and Society* 22(3):283-299.

Rose, N. and P. Miller. 1992. "Political Power beyond the State: Problematics of Government." *British Journal of Sociology* 43(2):173-205.

Russell, K.K. 1998. *The Color of Crime.* New York, NY: New York University Press.

Scraton, P. and J. Sim, eds. 1987. *Law, Order and the Authoritarian State.* Milton Keynes: Open University Press.

Simon, R. 1982. *Gramsci's Political Thought: An Introduction.* London: Lawrence and Wishart.

Smart, C. 1992. "Whiteness as Property." *Harvard Law Review* 106:1707-1727.

Solomos, J., B. Findlay, S. Jones, and P. Gilroy. 1982. "The Organic Crisis of British Capitalism and Race: The Experience of the Seventies." In *The Empire Strikes Back,* edited by The Center for Contemporary Cultural Studies. London. Hutchinson

Statistics Canada. 1996. "Public Perceptions of Crime and Criminal Justice." *Juristat* 11(1). Ottawa: Statistics Canada.

Steele, A. 1993. "The Myth of a Color-Blind Nation." *Capital University Law Review* 22:589-591.

Stenson, K. 1993. "Community Policing as a Governmental Technology." *Economy and Society* 22(3):373-389.

Sumner, C. 1979. *Reading Ideologies: An Investigation into the Marxist Theory of Ideology and Law.* London/New York: Academic Press.

Taylor, I. 1980. "Martyrdom and Surveillance: Ideological and Social Practices of Police in Canada in the 1980s." *Crime and Social Justice* 26:60-78.

The Report of the Commission on Systemic Racism in the Ontario Criminal Justice System. 1995. Queen's Printer for Ontario.

Toronto Life. 1994. "Guns and Roses: Georgina Leimonis Murder." 28(9):7.

Trotman, D. 1986. *Crime in Trinidad: Conflict and Control in Plantation Society.* Tennessee: University of Tennessee Press.

Van Dijk, T. 1993. *Elite Discourse and Racism.* London: Sage Publications.

Vincent, C.L. 1990. *Police Officer.* Ottawa, Canada: Carleton University Press.

Visano, L.A. 1994. "The Culture of Capital as Carceral: Conditions and Contradictions." In *Carceral Contexts: Readings in Control,* edited by K.R.E. McCormick. Toronto, Ontario: Canadian Scholars' Press Inc.

Weissberg, R. 1997. "White Racism: The Seductive Lure of an Unproven Theory." *Weekly Standard,* March 24.

Williams, Patricia. 1987. "Spirit-Murdering the Messenger: The Discourse of Finger-pointing as the Law's Response to Racism." *The University of Miami Law Review* 42(147):127-157.

Wilson, A. 1990. *Black-On-Black Violence*. New York: African World Info Systems.

Wilson, J.Q. 1992. "Crime, Race, and Values." *Society* 91 November/December.

Young, J. 1988. "Radical Criminology in Britain: The Emergence of a Competing Paradigm." *British Journal of Criminology* 28(2):289-313.

INDEX

A

Administrative Criminology, 2, 6-7, 9
African Canadian Legal Clinic, 74
Alderson, J., 86, 91
Alpert, G.P., 113
Althusser, L., 18, 19, 55-58, 110
Anti-Racism, 19, 27, 71
Asian Canadians, 99-101
Aubert, V., 58
Auletta, K., 2, 110

B

Baker, D., 17, 40, 51, 110-111
Balut, J.M., 49, 110
Barbados, 68
Barbalet, J.M., 66, 110
Barker, M., 9, 110
Barrett, M., 60, 110
Bayley, D.H., 110
Belgium, 69
Bennett, T., 57, 110
Black Action Defense Committee, 44
Black Communities, 4, 18-19, 23-24, 29, 35, 37-38, 87-89, 94-95
Black Crime Rate, 4-5
Black Crime, 3-5, 7-8, 26, 48-49, 52
Black Criminality, 2, 8, 50-51
Black Middle Class, 2
Boggs, C., 53, 110
Bridges, L., 96, 110
Britain, 69, 70, 85, 87, 97
Brown, L., 8, 27, 110
Burton, F., 11, 12, 13, 14, 17, 18, 19, 110

INDEX

C

Cain, M., 26, 110
Calliste, A., 68, 69, 111
Canada, 11, 12, 33, 38-39, 42-43, 62, 66-77, 80-81, 84-85, 87-88, 91, 94, 97, 99-100, 102, 106, 109
Canadian Association of Chiefs of Police, 22
Canadian Jewish Congress, 44
Canadian Legal System, 75-76
Caribbean Association, 69
Caribbean Immigration, 67
Carlen, P., 11-14, 17-19, 110
Cartwright, T., 11, 111
Cashmore, E., 9, 111, 113
Chinese Canadian National Council, 44
Clare Lewis Reoprt, 95
Clarke, M.J., 9, 15, 111, 113
Class, 40-41, 50, 52-53, 55-57, 59, 62-64, 67, 69, 71, 75-76, 82
Cloward, R., 7, 111
Cohen, A., 111
Cohen, P., 111
Cohen, S., 6, 20-21, 97, 111
Cohen, T., 75
Cole-Gittens Report, 12, 16, 18-26, 29, 31, 35, 86, 94-95, 97
Community Policing, 10-11, 20-24, 26, 31-33, 36, 38, 67, 84, 86-88, 90-92, 96-98
Cook, D., 66, 67, 111
Crime Statistics, 43, 48-49
Crime, 1-12, 17, 21, 23, 26, 28-29, 31-32, 34-37, 40-44, 46-52, 54, 56, 58, 60, 62, 64
Criminal Justice System, 31-33, 36, 40, 49, 52-53, 55-56, 66, 84, 94-95, 97
Criminal Justice, 1-3, 6,-10, 12, 19, 21, 22, 24, 28-29, 31-33, 36, 40, 49, 52-53, 55, 56
Criminality, 1-3, 7-8, 30
Criminalization, 1, 8-9, 11, 28, 40, 42, 44, 46, 48, 50, 52, 54, 56, 58, 60, 61-62, 64

INDEX

Criminology, 102
Critcher, C., 113
Critical Criminology, 2, 8-10
Critical Race Theory, 28
Cultural Racism, 49
Culture, 41, 46, 48, 50, 61

D

Denmark, 69
Deviance, 40, 50-52
Discourse Analysis, 16-19, 27
Discrimination, 17, 24, 28
Drugs, 89

E

Education, 4, 25, 29, 85
Employment Equity, 2, 44
Equality, 67, 75-76
Ericson, R.V., 85, 91, 99, 104, 111
Ethnic Relations Unit, 44
Ethnic Studies, 102, 104
Exclusion, 66-83

F

Fantino, J., 48, 49
Federal Multicultural Act, 87
Fielding, N., 8, 111
Findlay, B., 9, 116
Finland, 69
Fitzpatrick, P., 10, 111
Fleming, T.O., 111
Foucault, M., 13, 18-19, 26, 57, 59, 61, 110, 112
France, 69

G

Garland, D., 11, 29, 112
Gates, H.L.Jr., 112
Georges-Abeyie, D., 36, 112
Germany, 69, 72
Gilroy, P., 8-9, 50-51, 78, 80, 112, 116
Gordon, P., 91, 112
Gramsci, A., 29, 52, 54-56, 58-59, 110, 112-113, 116

H

Habermas, J., 15, 112
Haggerty, K.D., 85, 111
Hall, S., 8, 15-16, 41, 49, 55, 57, 62-63, 112-113
Harris, D.A., 113
Harris, R., 18, 28
Harrison Act, 62
Hawkins, D.F., 69-71, 113
Hegemony, 15, 17, 29-30, 52-55, 57-61, 64
Henry, S., 21, 113
Holland, 69
Hudson, B., 2-3, 6-9, 111, 113

I

Ideology, 40, 55-58, 60
Immigrants, 2, 44, 50, 62-63, 87
Immigration, 66-83
Inequality, 67, 75-76, 83
Institutional Racism, 46, 49

J

Jackson, P., 42, 45, 49, 52, 113
Jamaica, 68, 70, 73
James, J., 3, 17, 26, 113

Jefferson, T., 8-9, 15, 113
Jones, S., 9, 116
Junger, M., 2, 113

K

Kappeler, V.E., 113
Kobayashi, E., 42, 114

L

Lacan, J., 17-18
Law Enforcement, 31-35
Law, 41, 45, 47, 50, 52-53, 55-60, 66, 73-76, 80, 82-83
Lea, J., 9, 114
Lears, T.J.J., 52, 54, 114
Legal Hegemony, 58-60
Leighton, B., 28, 84, 97, 99, 115
Lemon, J.T., 42, 44, 114
Levi-Strauss, C., 18
Lewis, C., 8, 29, 114
Lister, R., 67, 79, 114
Lumley, B., 55, 113

M

Manning, P., 21, 114
Marchi, S., 71, 75, 80
Marginalization, 64
Marshall, T.H., 66-67, 114
Matas, D., 75
McCauglin, E., 9
McIntyre, C.C.L., 41, 114
Mclaughlin, E., 111, 113
McLennan, C., 111, 113
McLennan, G., 55
McNulty, E.W., 8, 114

Mensch, E., 59, 114
Merton, R., 7, 114
Metropolitan Toronto Police Association, 44
Metropolitan Toronto Police Services, 43-44, 92
Milivanovic, D., 113
Miller, P., 20, 88-89, 112, 114-115
Mills, C.W., 58, 115
Moore, M., 30
Mugging, 41, 49
Multiculturalism, 42
Murphy, C., 10, 115
Murry, C., 3, 115

N

Neo-Nazis, 72
New Racism, 9
Newman, K., 86
Noorani, R., 71-72, 82, 115
Normandeau, A., 28, 84, 97, 99, 115
Norway, 69
Nunn, K.B., 60, 115

O

Ohlin, L., 7, 111
Omi, M., 17, 115
Ontario Criminal Justice System, 12, 19, 21, 28-29
Ontario, 43-45, 76, 78-79, 84-85, 97-98

P

Parekh, B., 12, 115
Pauperism, 12
Penal Administration, 12
Pitman, W., 42
Police Brutality, 93

INDEX

Police Discretion, 24-26
Police Harassment, 92-93
Police Professionalism, 26
Police Racism, 49
Police Shootings, 99, 103, 108
Police, 2, 8-10, 20-26, 28-29, 84-98
Police-Community Relations, 42-44, 46, 48, 84, 87-88
Poor-Law Administration, 12
Poverty, 3-4, 41-42, 62, 66-67, 79
Pratt, M., 115
Prejudice, 17, 26-28
Prostitution, 89

R

Race Politics, 51
Race Relations Policy, 44
Race Relations, 99-100, 102, 107
Race, 1-11, 19, 24-29, 31-33, 35-37, 40-48, 51, 57-58, 60, 63, 66, 68, 70, 72-78, 80, 82-83, 99-102, 104-108
Racial Inequality, 17, 23, 24
Racial Prejudice, 17, 28
Racialized Crime, 49
Racializing Crime, 40, 42, 44, 46, 48, 50, 52, 54, 56, 58, 60, 62, 64
Racism, 1-5, 8-9, 11-12, 15, 17, 19, 21-29, 33, 36, 38, 66, 71, 75-80, 87, 96, 97, 100, 102-109
Radical Criminology, 2, 7-9
Reform Party, 71-73
Reiner, R., 3, 8, 20, 46, 115
Report of the Commission on systemic Racism in the Ontario Criminal Justice System, 28
Responsibilization Strategy, 11
Roach, L., 87
Roberts, B., 113
Robinson, V., 69, 115
Rock, P., 1, 3, 115
Rose, N., 13, 20, 88-89, 114-115

Royal Canadian Mounted Police, 73
Russell, K.K., 4-5, 40, 51-52, 92, 94, 115

S

Sadigh, S., 26, 110
Scraton, P., 20, 85, 112, 116
Sexual Morality, 16
Sim, J., 8, 112, 116
Simon, R., 53, 116
Sluder, R.D., 113
Smart, C., 116
Social Alienation, 41
Solomos, J., 9, 63, 116
Statistics Canada, 43, 84, 91, 116
Steele, A., 76, 116
Stenson, K., 20, 116
Storch, R., 86
Street Crime, 41, 50, 86
Sumner, C., 53, 59, 116
Sweden, 69
Switzerland, 69
Systemic Racism, 8, 12, 19, 21, 23, 28

T

Taylor, I., 20, 116
The Report of the Commission on Systemic Racism in the Ontario Criminal Justice System, 116
Thomas, R., 113
Toronto, 42-44, 47-49, 64, 88, 92, 98
Trinidad and Tobago, 70
Trotman, D., 41, 116

U

Underemployment, 64

INDEX

Unemployment, 41, 62, 64
United Nations, 70
United States, 38, 62, 85, 87, 97
Urban Politics, 99, 102, 104

V

Van Dijk, T., 22, 116
Vincent, C.L., 8, 116
Visano, L., 10, 52-54, 61, 116

W

Weissberg, R., 116
Welfare, 2, 3, 66, 77, 79
White Crime, 4-5
White Racism, 3-5
Williams, P., 28, 117
Willis, A., 8, 110
Wilson, A., 117
Wilson, J.Q., 3-5, 33, 117
Winant, H., 17, 115
World War II, 70
Wright, C., 71-72, 82, 115

Y

Young, J., 3, 7, 9, 26, 114, 117

Printed in the United States
52191LVS00005B/82-195